Your Rabbit
A Kid's Guide to Raising and Showing

NANCY SEARLE

Storey Publishing

The mission of Storey Publishing is to serve our customers by publishing practical information that encourages personal independence in harmony with the environment.

Cover and text design by Carol J. Jessop
Cover photograph by Positive Images, Jerry Howard
Production by Carol J. Jessop
Edited by Gwen W. Steege and Lorin A. Driggs
Illustrations by Carol J. Jessop, with the exception of those on pages 32, 36, 38, 40, 46, 49, and 53
Technical review by Terry E. Reed, D.V.M.
Indexed by Kathleen D. Bagioni

Printed in the United States by Versa Press
30 29 28 27 26 25 24

Library of Congress Cataloging-in-Publication Data

Searle, Nancy, 1945–
Your rabbit : a kid's guide to raising and showing.
p. cm.
Includes bibliographical references and index.
Summary: A handbook for raising rabbits for pets, for profit, or to show.
ISBN 978-0-88266-767-6 (pbk.)
1. Rabbits—Juvenile literature. [1. Rabbits.] 1. Title.
SF453.2.S43 1992 91-57949
636'.9322—dc20 CIP
AC

Contents

A Few Words to Parents / iv

Introduction / v

1
Choosing Your Breed of Rabbit / 1

2
Choosing Your Rabbit / 13

3
Handling Your Rabbit / 21

4
Rabbit Housing / 27

5
Feeding Your Rabbit / 47

6
Taking Care of Your Rabbit's Health / 59

7
Breeding, Birth, and Care of Newborns / 77

8
Showing Your Rabbit / 95

9
Marketing Your Rabbit / 111

10
Managing Your Rabbitry / 117

11
Activities for Young Rabbit Owners / 135

Glossary / 143

Index / 147

A Few Words to Parents

As a parent, you want to direct your children toward positive experiences that will help prepare them for an active, productive, and independent future. Caring for a rabbit can be a very special learning opportunity for your child. The role of caregiver can help your child develop a sense of responsibility and a sense of compassion for others.

As a youth project, rabbits are gaining in popularity for many reasons:

- Rabbits can be raised almost anywhere—they fit easily into most family settings.
- Raising rabbits gives kids lots of options, from a beginner's pet project to a breeding project and perhaps on to a small business venture.
- Rabbits are a good size animal for young children to work with. Young people are very capable of learning the skills necessary for a successful rabbit project.
- It doesn't take a lot of money to get started with rabbits—this project will fit into most family budgets.

Your personal interest and support will add greatly to your child's experience. Take time to work and learn together — enjoy!

Introduction

Rabbits and kids have a lot in common. Both come in a variety of sizes, shapes, colors, and personalities. Both can live happily in country, suburban, or city settings. Because rabbits are small, you will be able to take responsibility for most, if not all, aspects of a rabbit-raising venture. When you are just starting out, you needn't spend a lot of money on expensive equipment, either. Although proper equipment contributes to the success of a rabbit project, it is more important for equipment to be functional than fancy. If you start with only one, or even up to three, rabbits, you can keep expenses fairly low. As you gain experience and knowledge, you can expand the number of animals under your care.

If you enjoy animals, you will discover that rabbits offer a wide variety of opportunities to learn new skills and participate in new activities.

Rabbits Are Mammals, Too

Like people, rabbits are classified by scientists as *mammals.* All mammals

- Are warm-blooded
- Are nourished for a few weeks or months after birth on milk made by the mother's body

- Have hair
- Have more complex brains than other animals

Neither the jackrabbit nor the snowshoe rabbit is really a rabbit. Both of these animals are hares.

As scientists further break down the class of mammals, humans are grouped with other *primates* (including apes and monkeys), and rabbits are put into a special group called *lagomorphs,* which means *hare-like.* For a long time, rabbits were thought to belong to the rodent group, which includes other gnawing animals, such as rats, mice, and squirrels. Although rabbits are a lot like rodents, scientists saw enough differences to place rabbits in the group called lagomorphs. The

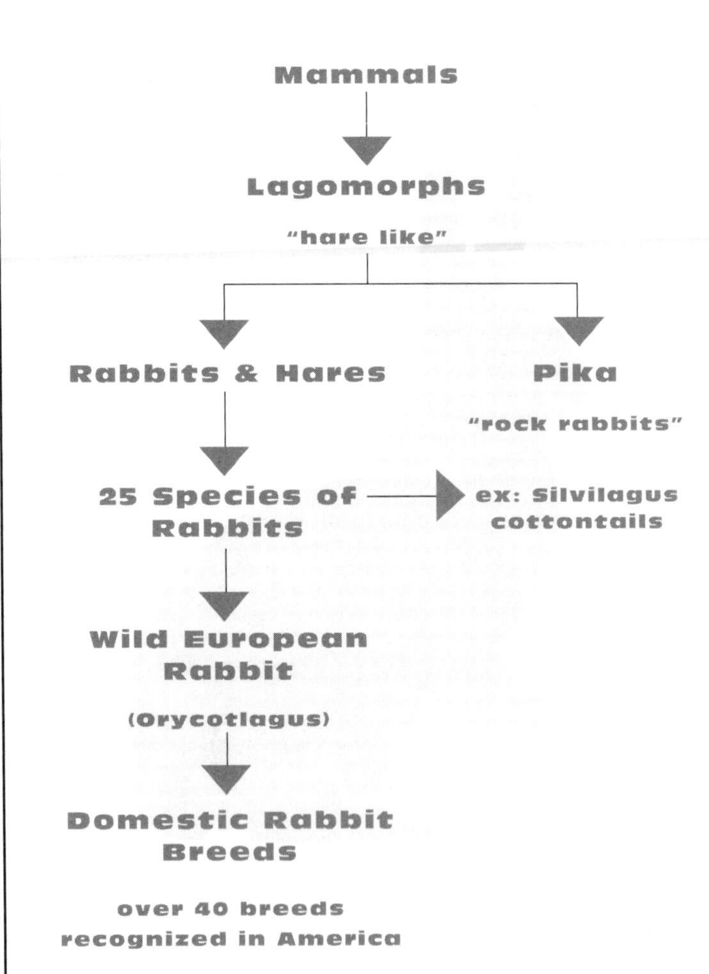

biggest difference between rodents and lagomorphs is their teeth. Rabbits have six front teeth (four on the top, two on the bottom). Rodents have only four front teeth (two on the top and two on the bottom).

Lagomorphs are subdivided into two families: 1) pikas and 2) rabbits and hares. Pikas are small, short-eared animals that live in rocky, mountain areas. Many people use the word *rabbit* and the word *hare* to mean the same animal. Although the two are closely related, you can tell them apart in these ways:

- Hares are usually larger than rabbits. Their legs are longer, their ears are longer, and they run with long, high leaps.
- Baby hares are born with fur, their eyes are open at birth, and they can move around shortly after they are born. Baby rabbits are born naked (without fur), their eyes are closed (they open at about 10 days), and they remain dependent on the mother for some time (5 to 8 weeks).

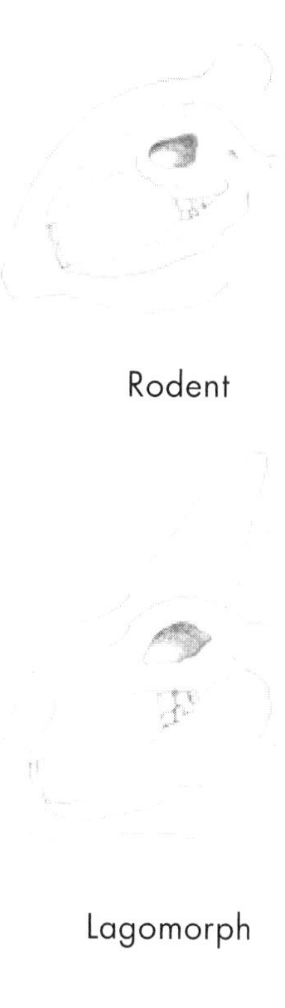

Smaller than hares, rabbits raise their young, which are born without fur, in a protected nest.

- Hares have their young in very simple nest areas. Baby rabbits must grow and develop within the safety of their nest, and so their mothers work harder to prepare a protected place.

Our Pet Rabbits

Breed. *A group of animals that have some of the same characteristics.*

Where does the tame rabbit that we raise fit into the picture? The rabbit and hare groups are divided further into various species. Worldwide, there are twenty-five species of rabbits. In North America, we see two of these species—wild and tame. Our wild rabbits include the Eastern Cottontail, the Desert Cottontail, and the Marsh Rabbit. Our tame rabbits are related to the wild European Rabbit. Although the wild European Rabbit does not live here in the wild, the *breeds* raised here as pets and for fur and meat are descended from wild European Rabbits.

A Little History

By studying fossils, scientists can tell that the rabbit has been around for about 30 to 40 million years. During all those years, the rabbit has changed very little. When humans arrived on the scene, wild rabbits became part of their diet. Exactly when people decided to raise rabbits rather than hunt them in the wild is unknown. We do know that early sailors took rabbits along on voyages because they were easy to feed and care for, and could be used as meat. These early sailors are given credit for introducing rabbits to some of the new places they explored.

Early European settlers probably brought rabbits with them to the New World, but since there were lots of wild rabbits and other wild game in America, the settlers didn't need to raise many rabbits. It was not until the early 1900s that tame rabbits were imported in any great numbers. The first real interest focused around the Belgian Hare (a breed of rabbit, not a true hare). Promoters led people to believe that raising Belgian Hares would make them rich. It didn't work out that way, however, and although some breeders were quite successful, many others lost money when they invested in these rabbits. Although the Belgian Hares didn't do all their promoters promised, however, they did greatly increase Americans' interest in rabbits.

Since 1900, other breeds have been imported and several American breeds have been developed. At the present time, more than forty breeds are recognized in North America by the American Rabbit Breeders Association (and over seventy-six breeds in the world).

How Are Rabbits Used?

Rabbits are just one of the many animals that have been *domesticated* (tamed) for people's use. If we think about the many ways that different animals have served people, we will find that rabbits have had an unusually broad range of uses.

***Domesticated**. Trained to being among humans and being used by humans.*

- *Pets.* Rabbits make wonderful, easy-to-care-for pets, even for those with little extra space.
- *Hobby.* Rabbits come in many breeds and colors. This variety makes them an interesting and challenging animal to raise and show. Sharing the experience with others makes the hobby of raising rabbits even more fun.
- *Meat.* Wild, and later tame, rabbits have provided meat for people throughout history.
- *Fur.* Rabbit furs have been used to make warm clothing.
- *Wool.* Angora rabbits produce wool (not fur). Angora wool is used to make knitted garments that

CHAPTER 1

Choosing Your Breed of Rabbit

Selecting a breed is probably the most important decision you will make for your rabbit project. Your choice of breed affects many things, such as the size of the cage you will have to build and when you should first breed young rabbits. Read as much as you can and seek out places where you can look at different breeds, so that you are better able to decide upon the one that best suits you. With over forty recognized breeds in this country to choose from, you will want to take some time to get familiar with a good number of them.

As soon as you start learning about breeds, you will also hear about *crossbred* rabbits. These are rabbits that have more than one breed in their family background. Although it's always easy to find crossbred rabbits, and they certainly are very appealing, I believe it is best to begin with *purebred* rabbits. Purebred rabbits may cost more than crossbred rabbits, but here are some of the many advantages that soon make up for the difference in price:

Crossbred. *Born of parents who are of different breeds.*

Purebred. *Born of parents who are of the same breed.*

- It costs just as much to house and feed a crossbred as it does a purebred.
- If you plan to breed and sell rabbits, purebred young bring a better sale price. You can therefore soon make up the difference in the initial cost of your *parent stock* (the mother and father).

■ If you want to show your rabbits—and showing is fun!—why not raise rabbits that will qualify to exhibit at a greater number of shows?

A good way to begin learning about the different breeds is to pay a small fee and become a youth member of the American Rabbit Breeders Association (A.R.B.A. For address, see page 140.) This organization has a lot to offer. Your membership will include a guidebook that has pictures and a written description of every breed. The American Rabbit Breeders Association also publishes a book called The *Standard of Perfection,* which can be very useful to you. The *Standard* gives a detailed, up-to-date description of each breed. This is a good book to own.

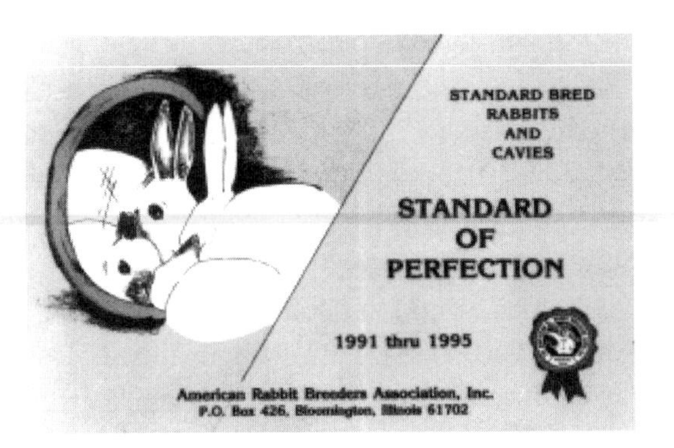

This book has pictures and descriptions of every rabbit breed.

Besides reading about the different breeds, find out where you can see rabbits in your area. If an agricultural fair is held in your community, be sure to check out the rabbit department. Local breeders will be sure to have animals on display. This is an opportunity to meet with breeders and to see several different breeds. An even larger number of breeds can usually be found at local rabbit shows that are held under the guidelines of the American Rabbit Breeders Association. These shows are called *sanctioned shows,* and they usually attract exhibitors from a fairly large area. If you become a youth member of the American Rabbit Breeders Association, you can find out about rabbit shows in your area by looking in the magazine that comes to you with your membership. Each issue includes a listing of upcoming sanctioned shows. Your membership also brings you a yearbook that lists all A.R.B.A. clubs and members. If you write a note to the club nearest you, you will find out about all the club's scheduled events, including shows.

Whether you attend a fair or a sanctioned show, try to make yourself just look and learn the first few times you go—don't buy. This won't be easy, because there will be lots of rabbits for sale and you are sure to see several that you would like to take home. Try to put off a purchase until you have found a breed that appeals to you *and* one that will also fill your needs as a breeder. Before buying, it's important to decide exactly *why* you are going to raise rabbits. Do you want to raise rabbits as pets or because you would like to make a hobby of breeding and /or showing rabbits; or are you interested in raising rabbits in order to sell their wool, fur, or meat? Since rabbit breeds have been developed for certain characteristics, you will find some breeds are more suited to one or more purposes than to others.

Good Breeds for Pets

An increasing number of people keep rabbits as pets. As pets, rabbits are relatively inexpensive, they can be kept indoors or out, they make no noise, they have few veterinary needs (including no vaccination), their initial cost is low, and daily care is not demanding. These features make rabbits ideal pets for busy, modern families. Any breed may be kept as a pet, but if you want to raise rabbits primarily to sell as pet rabbits, you are wise to consider some of the smaller breeds.

Netherland Dwarf. If you begin showing your animals you will learn that among the many rules are standard weights for each breed. The Netherland Dwarf, for instance, is the smallest rabbit breed and, to be shown, they may be no more than 2½ pounds when they are mature. This breed is known for its round body, broad head, large, bold eyes, and short ears. Because it is small and has such a bright appearance, it makes a very appealing pet. The Netherland Dwarf comes in a wide variety of colors—over thirty different varieties are recognized. This breed has small litters (two to three young in an average litter), so if you want to breed Netherland Dwarfs, you won't have as many to sell as you would if you were raising most other breeds.

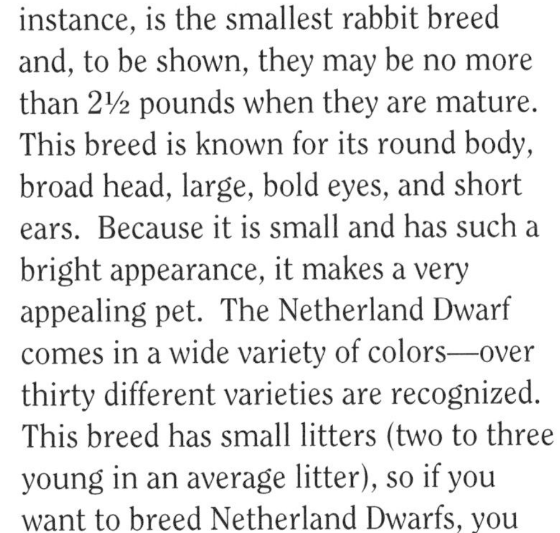

Netherland Dwarf

Dutch. The distinctive markings of the Dutch make them very attractive pets. Their small size (3½ to 5½ pounds), plus their mild personality, make them a good choice for young children. Dutch are available in six color varieties. Not all the young in a litter will be marked in the way needed to make them of show quality.

PAMELA T. BERNARDINI

Dutch

Mini Lop. Their outgoing personality and cute lop-ears make the Mini Lop a popular choice. The breed comes in many colors. You'll find them grouped in two show categories: *colored pattern* or *broken pattern,* which includes any recognized color combined with white. Although the Mini Lop is no longer the smallest lop-eared breed, its top weight of 6½ pounds makes it a good choice even for families without a lot of space, and it is a breed that is easy for beginners to raise.

Mini Lop

Holland Lop. Smaller than the Mini, the Holland Lop weighs 3 to 4 pounds when it is mature. You can tell by its broad, bold head and small, compact body that it is closely related to the Netherland Dwarf. Like other lops, the Holland comes in many colors. At shows, it will be placed in either a colored or broken category.

PAMELA T. BERNARDINI

Holland Lop

Raising Rabbits as a Hobby

It is very common for people who have raised a rabbit or two as a pet to become interested in taking on the hobby of breeding rabbits to sell them. All breeds are suitable for this activity, and the most important reason to decide which breed to raise is simply to choose one that appeals to you. You may have to take a second or third choice depending on these other factors that must also influence your decision:

- *Availability.* Is your chosen breed raised in your area? If you select a breed that is popular locally, you will have a greater choice of quality animals.
- *Cost of animal.* You will find that some breeds command higher prices than others.
- *Cost of care.* Other factors, such as breed size, also affect overall costs, since larger breeds need larger housing and consume more feed.
- *Amount of care.* Angoras, which must be groomed daily, take more time than other breeds.
- *Ease of raising.* Some breeds, such as Netherland Dwarfs and Holland Lops, can be difficult to raise.

Angora Breeds

In recent years, natural fibers and hand-crafted items have become very popular. This has brought about a new appreciation for the Angora rabbit breeds. Soft and warm, Angora wool is obtained by pulling the loose hair from the mature coat. Because you are really just helping the natural shedding process, this hand plucking does not hurt the animal. The plucked wool can then be spun into yarn. Clothing made from Angora wool brings a very high price. Although Angora rabbits require some special management (they need to be groomed often), raising them and selling their wool is an excellent project for young people.

English Angora. This is the fanciest of the Angora breeds. A relatively small animal (5 to 7½ pounds), it is known for its luxuriously soft wool. The breed includes two color varieties, white and colored, of which there are approximately thirty different shades.

French Angora. Larger than the English Angora, the French weighs in at 7½ to 10½ pounds. The English and French differ in other ways, as well:

- French wool has a coarser texture than English.
- French has normal fur up to the first joint on its feet and legs, whereas the English has wool to the ends of its feet.
- The head of the French is plain, whereas the English has long decorative wool on its ears and face. This longer wool on the head of the English Angora is called *furnishings.*

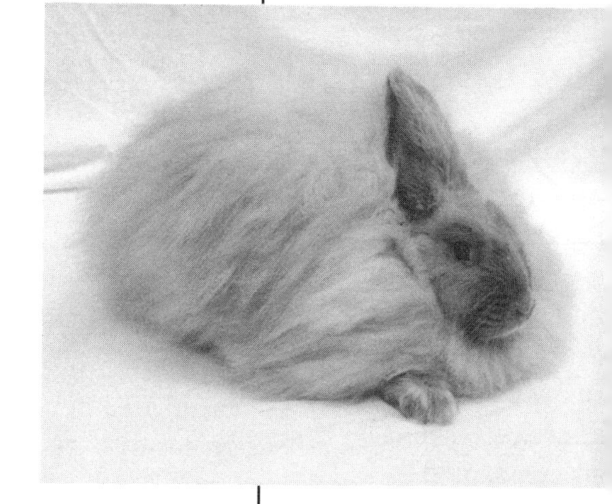

French Angora

Giant Angora. This is the largest of the Angora breeds. If you show them, mature animals must be at least 8½ pounds. The Giant Angora comes only in white. Like the English, the Giant has wool on all parts of its body.

Furnishing. *The long decorative wool on the head of an English Angora.*

Satin Angora. The unique feature of the Satin Angora is that its wool is similar to that of Satin fur rabbits (see page 12). Satin Angora wool has extra sheen because, like the Satin fur rabbit, its hair shaft is more transparent than that in normal fur. Like English and French Angoras, the Satin comes in white and colored varieties.

Two other rabbit breeds have wool, instead of normal fur—the Jersey Woolly and the American Fuzzy Lop. These are small breeds. Due to their small size, they produce less wool and so they are more suitable for pets or show than for wool production.

NATIONAL ANGORA RABBIT BREEDERS CLUB

Satin Angora

...aising ...abbits for ...eat

...hen you raise rabbits ...r meat, you will not ...ink of them as pets. ...owever, all rabbits in ...ur care should

- Be treated with kindness and respect
- Have clean housing with adequate room that protects them from harsh weather conditions
- Receive appropriate nourishment and health care
- Be protected from danger, such as wild animals

Meat Breeds

If you are interested in meat production, consider those breeds that are termed "medium-sized breeds." These weigh about 9 to 12 pounds when they are mature. Although not all medium-sized breeds are ideal for meat production, the following popular breeds are worth looking into.

New Zealand. This breed has long been the top contender for the Number One meat-producing breed. Although it comes in three color varieties—red, white, and black—it is the white that has proven most popular for serious rabbit-meat-producing businesses. The New Zealand is known for its full, well-muscled body type and for its ability to grow market-ready fryers (4 to 5 pounds live weight) by 8 weeks of age.

Californian. Running neck and neck for first place with the New Zealand, the Californian is another outstanding meat-type breed. This breed comes in only one color—a white rabbit with black coloring on the feet, tail, ears, and nose.

PAMELA T. BERNARDINI

New Zealand

PAMELA T. BERNARDINI

Californian

Champagne D'Argent. One of the oldest rabbit breeds, this silver-colored rabbit is born completely black and gradually turns silver as it matures.

CHAMPAGNE D'ARGENT RABBIT FEDERATION

Champagne D'Argent

Palomino. This is another breed that combines an attractive color with a body type that is well suited for meat production. Developed in the United States, rabbits in the breed come in two colors, golden and lynx.

PAMELA T. BERNARDINI

Palomino

Florida White. Although the Florida White is not in the 9 to 12 pound range, it is a good meat-type rabbit. This 4 to 6 pound rabbit was an American creation. Its smaller size makes it especially appropriate for young rabbit raisers.

PAMELA T. BERNARDINI

Florida White

Fur Breeds

Raising rabbits just for their fur is not very practical, unless you want to try home tanning or send your *pelts* (skin or fur of an animal) out to be processed. Fur markets are few and far between, and fur production is more suited to large rabbitries with many, many rabbits than to small rabbitries that people have at their homes. Most of the breeds that you might choose for fur production are also quite good for meat production. The Rex is the most valuable fur breed, but white-furred rabbits are quite desirable for fur production, because their pelts can be easily dyed to a variety of colors. The New Zealand and the Californian breeds are excellent choices. Even though the Californian has black on its feet, tail, ears, and nose, it is considered a

Pelt. *Skin or fur of an animal.*

lutation. *Sudden, nexpected change in haracteristics that can e passed on to off- pring.*

nherit. *Receive a pecific characteristic om a parent or other ncestor.*

white rabbit to furriers because the usable portion of the pelt is white. The fur breeds mentioned below are unique because of special features of their fur.

Rex. Because of its very short, plush fur, the Rex is the original "Velveteen Rabbit." The Rex came about when an unusual fur appeared in a litter of normally furred rabbits. An unexpected change such as this is called a *mutation.* The Rex breed was created by mating two rabbits that had this mutation. Their offspring then *inherited* the plush fur. This breed is available in fifteen different color varieties, including white.

Satin. This is another example of a separate breed being developed from animals with a fur mutation. In the fur on Satins, the hair shaft is transparent, which allows more light to pass through and gives the Satin breed its special shine or sheen. Satins come in white and nine other color varieties.

PAMELA T. BERNARDINI

Satin

Choosing Your Rabbit

CHAPTER 2

After you have decided the breed of rabbit you are going to raise, your next steps are to learn how to make a good selection when you choose your rabbit, make some decisions about your own needs and preferences, locate breeders with stock for sale, and visit a rabbitry and choose your rabbit.

Judging the Breed

Although different breeds have slightly different characteristics, those characteristics all appear on the same basic rabbit body. If you read that your breed should have a large, deep loin, you need to know where the loin is. If it is a disqualification for your breed to have a dewlap, you need to know where to look for a dewlap. Taking some time to study the drawing on page 14 will help you make the best possible choice.

Recognizing a Healthy Rabbit

Good health is the most important quality to consider when you select your stock. What does a healthy rabbit look like? Look for the following features:

Eyes. The eyes are bright, with no discharge and no spots or cloudiness.

Ears. The ears look clean inside. A brown, crusty appearance could indicate ear mites.

Nose. The nose is clean and dry, with no discharges that might indicate a cold.

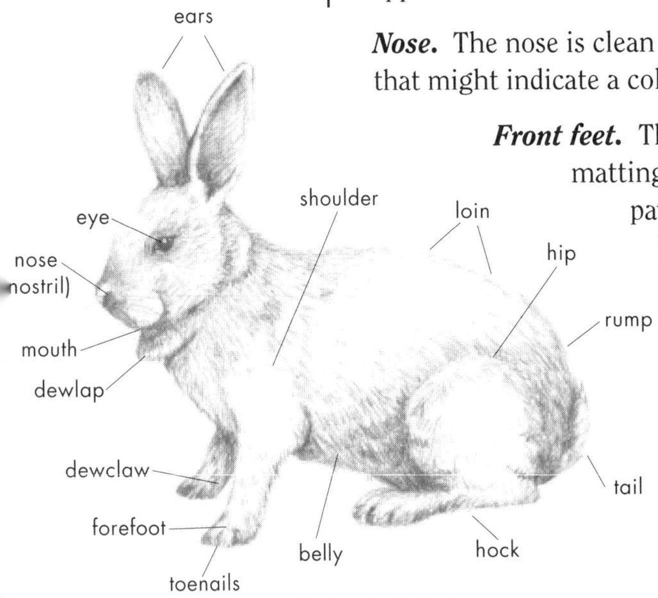

Front feet. These are clean. A crusty matting on the inside of the front paws indicates that the rabbit has been wiping a runny nose, and may have a cold.

Hind feet. The bottoms of the hind feet are well furred. Bare or sore-looking spots can indicate the beginnings of sore hocks.

Teeth. The front teeth line up correctly, with the front top two teeth slightly overlapping the bottom ones.

The parts of a rabbit

General condition. The rabbit's fur is clean. Its body feels smooth and firm, not boney.

Rear end. Area at the base of the rabbit's tail should be clean, with no manure sticking to the fur.

Characteristics of the Breed

Body type. *The proper size and shape for a particular breed.*

Each rabbit breed has distinctive features that make it different from all other breeds. These features include special colors, markings, *body type* (the proper size and shape for its breed), fur type, and weight. The more you know about the special characteristics of your breed, the better job you will do of picking out a good one. Detailed descriptions of breeds are found in the *Standard of Perfection,* as well as in printed information produced by breed specialty clubs.

Eliminations or Disqualifications

If you are purchasing purebred rabbits, you will want them to be free from any conditions that would prevent them from winning prizes at a rabbit show or even from being accepted for registration (pages 97–100). Many of the health-related conditions mentioned above are *eliminations* or *disqualifications* in all breeds. An elimination is a condition that is not permanent. Ear mites, sore hocks, and broken teeth are examples of eliminations. Disqualifications are considered to be permanent conditions. Some general disqualifications include blindness, missing toenails, bad teeth, and torn ears. The A.R.B.A. *Standard of Perfection* contains a complete listing of eliminations and disqualifications.

Trio. *A group of rabbits consisting of one buck and two does.*

Buck. *Male rabbit.*

Doe. *Female rabbit.*

Your Needs and Preferences

How Many Rabbits Do You Want?

Are you looking for one rabbit, or several? Many youth breeders start with one rabbit. This allows them to experience both the fun and the work that goes with rabbit ownership. If they enjoy caring for their rabbit, they can get more rabbits as time goes on.

Some new rabbit owners choose to start with more than one rabbit. A good number to begin with is three rabbits, consisting of one *buck* and two *does*, known as a *trio*. A trio of rabbits is a reasonable number for a beginner to care for, and they can become the start of your own breeding program. If you purchase a trio, you will want to buy animals that are not too closely related. Ask the breeder to help you select rabbits that can be used to start a breeding program. Also, look at the animals' pedigrees to find out whether they are closely related. The rabbits in your trio may have some relatives in common, but their pedigrees should show some differences in their family trees as well.

Whatever number of rabbits you begin with, remember that each rabbit will need its own cage. Don't purchase more animals than you are able to care for properly.

Your trio should not be brother and sisters. Three rabbits from the same litter are too closely related to be used in a breeding program.

What age rabbit do you want?

Baby rabbits have a better start in life if they remain with their mother until they are 5 to 8 weeks of age. Be sure your new rabbits are weaned from their mother (see pages 92–93) before you move them to a new home.

When you invest in rabbits, you want to enjoy them for as long as possible. Rabbits live an average of 6 to 8 years. For breeding purposes, rabbits are most productive between 6 months and 3 years of age.

2–3 months. Look first at rabbits that are 2 to 3 months of age. Young rabbits are cute and fun. You will enjoy caring for them and watching them grow. Younger animals are also smaller, so it will be easier for you to learn to handle them.

4–5 months. As cute as baby rabbits are, you can't always be sure exactly how they will look when they grow up. One advantage to purchasing slightly older rabbits is that you can get a better idea of what they will look like when they mature. When you select a rabbit that is 4 to 5 months old, you will still have a young animal, but you should also have a better idea of how the animal will look as an adult.

If you plan to breed your rabbits, it is best not to buy animals that are over 2 years of age.

6 months and older. When a rabbit is 6 months old, you can get a pretty accurate picture of what type of adult it will be. If you intend to breed it, an older rabbit will be ready for breeding sooner. However, if you purchase an animal at this age, you miss the fun of seeing it grow.

How much can you spend?

Rabbit prices vary by breed, age, availability, and popularity. They can range from $10 to over $100 per animal! Because so many factors affect the price, it is difficult to give an average. However, if you are buying a purebred rabbit of one of the common breeds, you should be able to find a good-quality animal for between $15 and $35. When you contact breeders, ask how much their rabbits cost. When you choose a breeder, be sure his or her prices fall within your budget. Some breeders may offer special prices if they know you are buying a rabbit for a 4-H, Scout, or other youth project. When you inquire about price, make sure the purchase price includes the *pedigree*.

Pedigree. *Written record of an animal's ancestors, going back at least three generations.*

Locating Breeders

To locate breeders with stock for sale, attend shows or work through breed or 4-H clubs.

Shows

Attending rabbit shows is a good way to find breeders. Shows also give you a chance to watch your breed being judged and to begin to gather specific information about the characteristics of that breed, so that you will recognize a good representative of the breed.

Shows are good places to contact breeders whose rabbitries you might visit at a later date when you are ready to purchase your rabbit. You'll also find rabbits for sale at shows. Don't let the excitement of the show cause you to rush into making a decision. Before you buy a rabbit at a show, take the time you need to be sure the rabbit is the right one for you, as described in the previous section. Be sure to talk to the breeder, and ask the same questions you would ask if you were buying from a rabbitry.

Show Courtesy

If you attend a show to learn about rabbits, please heed this advice. When the breed you are interested in is being judged, watch and listen. This is not the time for conversation. Judges have a job to do, and shouldn't be distracted. Breeders have to concentrate on getting their animals to the proper place at the proper time, and they also want to hear the judge's comments about their animals. Breeders and judges will be a lot more interested in answering your questions if you save them for a less hectic time.

Breed Clubs

Each rabbit breed has its own fan club of sorts, called a specialty club. One way to locate breeders is to contact the specialty club that represents your breed. The A.R.B.A. yearbook lists each breed club and the address of the club's secretary, who can give you the names and addresses of breeders closest to you. Write or call those breeders to find out if they have stock for sale. You'll increase your chances of getting a reply by mail if you write to several breeders and enclose a pre-addressed, pre-stamped envelope for replies.

4-H Clubs

Another network for locating breeders is your local 4-H community. Rabbit projects are popular in the 4-H program, and club members and leaders are likely to know of many local rabbit breeders. 4-H is the youth division of the Cooperative Extension Service, and there is a Cooperative Extension office in every county. Your local 4-H office can help you contact the 4-H rabbit project volunteer nearest to you. 4-H volunteers are well-known for being especially helpful.

At the Rabbitry

When you find a breeder who has stock available, plan a day when you can visit. Make an appointment, and do your best to arrive on time.

Visiting a rabbitry can be pretty exciting! You don't want to rush into making a decision, so take some time to look around. Ask the breeder to give you a brief tour of the rabbitry. You can learn a lot from how other people do things. Along with information about the general care of rabbits, the breeder also possesses in-depth knowledge about your breed. Invest some time in asking questions—one of the best ways to learn.

General Conditions

Before you look at individual animals, you should feel comfortable with what you see throughout the rabbitry. If you do not like what you see, this is probably not a good place to purchase your rabbits. Do remember, however, that a rabbitry does not have to be fancy and new to meet the needs of the rabbits. If the rabbits' cages are secure and clean, and if the rabbits are healthy and well fed, the breeder is doing a good job of providing for their welfare. Once the rabbitry passes your inspection, you can begin to look for that special animal that will be just right for you.

Making Your Selection

The breeder's knowledge and experience can be very helpful to you, if he or she knows what you're looking for.

- Do you want a buck or a doe?
- Are you planning to show the animal?
- What age rabbit are you looking for?
- About how much can you spend?

There may be several rabbits to choose from, so ask the breeder to remove the prospects from their cages so you can get a closer look. Placing the rabbits on a table or similar surface will give you a chance to see them next to each other and to compare them as to size, markings, body type, and so on. Having several rabbits together can also be confusing, so enlist grown-up helpers and the breeder to keep the situation under control. Ask which rabbit the breeder thinks is the best, and why. Use the information you've learned as well. Many of the conditions you'll want to know about require handling the rabbit. If you are not an experienced handler, ask the breeder to help by checking the rabbit's teeth, sex, toenails, and so on. If a rabbit you are interested in is young, ask to see its mother and father. This will give you an idea of what the rabbit should look like at maturity. Ask to see written records for the parents that indicate how well they have performed. These include the parents' pedigrees, plus records that provide information such as how many young the mother has raised, how many young the buck has sired, and perhaps information on how well the parents have placed at rabbit shows.

Things to Look for at the Rabbitry:

- Is the rabbitry clean?
- Do the rabbits look healthy?
- Is clean water and food available to the animals?
- Are the rabbits properly housed?

If you have examined the animal carefully and it has passed your inspection, if the parent stock looks good, if it meets your criteria, and if you and the breeder agree on price, you have found your rabbit!

inal ecision to uy

on't make a final ecision until you are ıre that the animal

- Is healthy
- Has no eliminations or disqualifications
- Meets the size, color, and other characteristics required of its breed
- Meets your own criteria, including price

Final Purchase Details

Before you head home, you still have details to attend to.

Payment. You need to pay for your rabbit.

Pedigree papers. If you are purchasing a purebred rabbit, the price should include pedigree papers. An organized breeder will have pedigree papers ready. Less organized breeders may choose to mail the papers to you at a later date. (See pages 129–31 for more about pedigree and registration.)

Food. Be sure to ask the breeder what kind of food the rabbit is currently receiving. How much is it being fed? How often is it fed? This information will enable you to help the rabbit make an easier adjustment to its new home. If you plan to use a different feed from that used by the breeder, you will want to plan a gradual change for the rabbit. Ask the breeder to supply you with a small bag—about 1 pound—of his feed. Give this feed to the rabbit for the first day or two. Then begin to mix the breeder's feed with your feed, gradually making a complete changeover to the new feed.

Guarantee. Ask about the breeder's policy concerning problems that may arise with your new rabbit. Most breeders will guarantee the health of their rabbits for at least 2 weeks. After 2 weeks, it is difficult to know whether a problem started at the breeder's rabbitry or was acquired after the rabbit was sold.

Handling Your Rabbit

If you are attracted to rabbits, you will find it hard not to want to pick one up and hold it! As simple and as basic as this sounds, there are some definite dos and don'ts when it comes to handling rabbits. As you practice handling your rabbit, your skill will increase, and repeated gentle handling will also make your rabbit easier to handle.

Being picked up can be a scary experience for a rabbit. When you lift a rabbit, you are taking away its most effective method of defense—its ability to run away from danger. In the wild, running away is how a rabbit protects itself. Rabbits need to have all four feet on something solid in order to run, so it's easy to imagine that removing this security could make a rabbit nervous.

If your rabbit is frightened, it will try to run away. A rabbit's toenails help it grip surfaces, enabling it to run faster. Sometimes when you lift the rabbit, you become the only surface the rabbit has to grip. This often results in scratches for the handler and a panic situation for both the handler and the rabbit. Many new rabbit owners become discouraged when their rabbit scratches. Remember, your rabbit is scared—it is not mad at you. You can make things easier on yourself by wearing a long-sleeved shirt.

Some breeds handle more easily and respond better than others. Most enjoy being stroked, especially when they are young. Remember always to talk calmly and quietly. Loud noises easily startle rabbits. Relax and enjoy your rabbit.

Getting Acquainted

A new rabbit has many adjustments to make. Give your rabbit a few days to get used to you and its new surroundings before you handle it a lot.

The more time you can spend working with your rabbit, the better. Learning how to keep your rabbit from being afraid will help both of you enjoy handling.

Get to know your rabbit in a setting where both of you can be comfortable. A good place to get acquainted is at a picnic table covered with a rug, towel, or some covering that will give the rabbit secure, not slippery, footing. Your rabbit will be able to move safely around on the table, and you can safely visit and pet your rabbit without having to lift it. You may need to ask an adult or more experienced friend to help by removing the rabbit from the cage and bringing it to the table. Sometimes getting a rabbit out of its cage can be a challenge. Rabbits will usually not jump off a table. In spite of this, *never leave the rabbit unattended when it is out of its cage.*

Once your rabbit seems comfortable on the table, this is a good location to begin practicing picking it up, because it offers a handy surface to set the rabbit safely back onto. This is much safer than risking a possible fall to the ground.

Picking Up Your Rabbit

The best way to pick up a rabbit is to place one hand under it, just behind its front legs. Place your other hand under the animal's rump. Lift with the hand that is by the front legs and support the animal's weight with your other hand. Place the animal next to your body, with it head directed toward the corner formed by your elbow. Your lifting arm and your body now support the rabbit—just like tucking a football against you for that long run. Your other hand is free and can rest on the rabbit's back for extra security. Place the animal gently back on the table and repeat this lift.

A "football hold" is a secure way to carry your rabbit.

It is best to practice handling skills often, but for short periods of time—about 10 to 15 minutes is all it takes. Have a short practice session each day until you and your rabbit are comfortable with each other.

Once you feel at ease, begin to move around while holding your rabbit. You may need to steady and secure your animal with your free hand until your rabbit gets used to being carried. At first, just walk around the table, so if your rabbit becomes frightened you have a safe place nearby to set it down promptly.

If you need extra control, you can pick up your rabbit with one hand by grasping the loose skin at the nape of its neck, while your other hand provides support under its rump. This method can damage the fur and flesh over the rabbit's back, however, and it is especially harmful to the more delicate coats of Rex and Satin rabbits. Once you have lifted the rabbit, move it close to your body where it will feel more secure. After it is tucked in "football style," you can use a one- or two-hand carry. The rabbit's behavior will help you decide whether to use one or both hands.

Sometimes an overactive bunny will struggle and get out of control. When this happens, drop to one knee as you work to quiet your animal. Lowering yourself to one knee lessens the distance the rabbit has to fall. You can also easily set the animal on the ground from this position, if necessary. After a short rest on the ground, carefully and securely lift the rabbit again. Even the most mild-mannered rabbit can have a bad day, so be prepared to handle any situation in a calm, controlled manner.

If your rabbit starts to struggle when you are holding it, drop to one knee.

Turning Your Rabbit Over

After you master lifting and carrying your rabbit, you will want to learn how to turn it over. You'll need to get it in this position in order to better observe many important things about it, such as its teeth, its sex, and the presence of toenails. You can't see everything about a rabbit by just looking at the top!

Turning a rabbit over puts the animal into a very unnatural position. First of all, its feet are off the ground so it cannot run away, and secondly, to make matters worse from its point of view, its underside is now exposed. A wild rabbit in this position is probably one that is about to be eaten by a predator! Your rabbit has good reasons to resist this type of handling, so you have to be especially careful and patient.

A good place to practice will be back at the table where you first learned how to carry your rabbit. Again, you will want to have a rug on the table, and you should remember to wear that long-sleeved shirt.

To turn the rabbit, you will use one hand to control the head and the other hand to control and support the hindquarters. Place the hand that holds the rabbit's head so that you are holding its ears down against its back, while you reach around the base of the head. If you prefer, you can place your index finger between the base of the rabbit's ears and then wrap your other fingers around toward its jaw. With your other hand, cradle the rump. Now that your hands are in place, lift with the hand that is on the head and at the same time roll the animal's hindquarters toward you. Try to do this movement in a smooth, unhurried manner.

If your rabbit cooperated, you will now be able to let the table support its hindquarters so that the hand that

was holding the rump is free to check other things, such as the teeth, toenails, and so on. If your rabbit fights against this procedure, it is its way of showing that this is not its first choice of positions. Keep trying, but be sure to do so in a place where the rabbit hasn't far to fall, and try to support it securely. Your rabbit can injure itself if it struggles and falls.

Part of the trick to successfully turning your rabbit over is being able to grasp its head so it cannot wiggle away. Small hands and big rabbits make this difficult, so you may want to begin practicing on small, young rabbits. A way to gain some extra control is to grasp the lower portion of the loin, instead of cradling the rump. Another approach is to pick up your rabbit and allow it to rest against the front of you instead of tucking it against your side. The animal will face upward with its feet against you. When it is in this upright position, place one hand on its head, as suggested above, and keep your other hand supporting its hindquarters. Now, if you slowly bend forward at your waist, you can lower the animal to the table, where it should arrive in the proper turned-over position.

If you need to turn your rabbit over for a closer look and no table is handy, you can let the animal rest on your forearm instead of on the table. When you have the rabbit in this position, you get better control over the hindquarters, because they are tucked between your elbow and your body. This is even easier if you sit in a chair. You can use your legs instead of the table to support your rabbit. In fact, you may find that when you are seated, you can hold your rabbit more securely for grooming, trimming toenails, and so on.

You can hold your rabbit against your chest, and then bend forward to lower it safely on a table.

Successful Handling Pays Off

The time you spend learning proper handling skills will pay off in many areas of your rabbit project. If you show your rabbits, animals that are used to being handled will generally perform better on the show table. They will also sell better, since you will be able to show them off to their best advantage, and prospective buyers will be able to get a better look at them. And rabbitry chores are easier if your animals cooperate during cage cleaning, grooming, and breeding.

Once you gain some confidence handling your rabbits, look into entering a rabbit showmanship contest. This competition measures your ability to handle your rabbit. It also tests your knowledge of rabbits and judges the health and condition of your rabbit. Showmanship is a challenging way to test your skills against other youth rabbit raisers. (For more on showing your rabbits, see pages 95–110.)

Rabbit Housing

CHAPTER 4

Rabbit housing—called a *hutch*—can be the largest single expense in your rabbit project. Whether you buy or build, you must be sure that your housing meets the needs of your animals. Proper housing can contribute greatly to the health and happiness of your rabbits.

Hutch. *Rabbit housing.*

Adequate Space

Your rabbit will spend most of its time in its hutch, and it will need space to move around, as well as space for feeding equipment. When you are planning how much space you need, a general rule is to allow 1 square foot of space for each pound your rabbit weighs. A cage 2 feet wide and 3 feet long is 6 square feet, and would be perfect for a 6-pound Mini Lop. Find out how much a mature animal of your rabbit's breed weighs, and use this as a guide to plan the size of the hutch you need. Young rabbits and bucks (males) may do fine with slightly less space, but does (females) that will be having *litters* (baby bunnies) will need the whole amount.

After you have figured the width and length of the cage, you must decide how high it should be. Most cages are 18 inches high, although small breeds, like the Netherland Dwarf, need a height of only 12 to 14 inches.

Requirements for Good Rabbit Housing

- Provide adequate space
- Protect your rabbits from the weather
- Protect your rabbits from other animals
- Be easy to clean

Litter. *Baby rabbits born in one birth.*

How to Figure Square Feet

Multiply the length of the hutch times the width of the hutch. Example: 2 feet wide x 3 feet long = 6 square feet

Protection from Weather

Many youth rabbit raisers have their rabbitries outdoors. I believe that, with proper care, rabbits have fewer health problems when raised outdoors. However, you must be able to adjust the housing to changes in weather. Where you live determines what weather conditions you will have to deal with. Rabbits are most comfortable if the temperatures are between 50° and 69°F. In hot climates, locate the cage in the shade, and make sure it has lots of air circulation around it. In cold weather, cages can be located in a sunny area. You should have a way to keep the wind off your rabbits. Some rabbitries have removable wooden panels that can be attached to the hutch, and some use plastic sheeting to enclose cages during the cold weather. In hot weather or cold, your outdoor cages should have a secure roof to keep out rain and snow.

Wrap plastic sheeting around three sides of your hutch, and nail a separate flap of plastic over the front.

Protection from Other Animals

A sturdy, well-built hutch will contribute to the safety of your rabbits. Most cages are built on legs or hung above the ground—which also makes the cages safer from dogs, cats, rodents, raccoons, and opossums. Keeping your cages in a fenced area or indoors is even better, but this may be impossible when you are just getting started.

Easy-to-Clean Cage

Clean cages are important to the health of your rabbits. If your cages are easy to clean, you will be able to do a better job of caring for your rabbits. Cages should have wire floors that allow droppings to fall through to a tray or to the ground. Choose all-wire cages or wood-framed cages that do not have areas where manure can pile up. Plan your cages so you can easily reach into them and get around them to clean all parts.

If You Have Second-Hand Cages

Many rabbit raisers begin with used cages. This is usually a less expensive way to get started, but be sure to do the following before you use second-hand equipment:

- Check carefully for holes, rusted wire, chewed wooden areas, and sharp edges. Repair any problems *before* you introduce your rabbits to their new home.
- Carefully clean and disinfect the cage before you use it. First, scrape off any old manure and hair. Then, wash the entire cage in a chlorine-bleach-

and-water solution. (You can make this solution by mixing 1 part household bleach with 5 parts water.) Set the cage in a sunny spot and allow the cage to dry thoroughly before you put a rabbit into it.

Sources of Rabbit Cages

- Rabbitry supply companies
- Farm supply stores
- Local rabbitry suppliers
- Pet shops

Where to Buy Cages

If you are lucky enough to begin with a new cage, you will want to purchase the best possible home for your rabbit. Where do you shop for rabbit cages? There are several choices, and some are better than others.

Mail-Order Rabbitry Supply Companies

These companies specialize in cages and equipment for rabbit breeders. They generally offer the best quality, selection, and price. Most rabbitry supply companies will send you a free catalog if you request one. A list of several suppliers is included at the back of this book.

When you decide to order a cage by mail, be sure to read the catalog to find out if the supplier gives you a choice of buying the cage assembled or unassembled. Assembled cages are more expensive and also cost more to ship. If you choose to save money and buy an unassembled cage, it will arrive as a package of pre-cut pieces that you will put together. When you assemble your rabbit cage, you'll need the use of a tool called *J-clip pliers.* (See page 37 for a picture of J-clips and J-clip pliers.) If you have one or just a few cages to assemble, you may be able to borrow J-clip pliers from a local rabbit raiser. If you are going to assemble lots of cages, you will want to purchase one of these tools.

Farm-Supply Stores

The same store where you buy your feed may also sell cages. You will probably find that prices are fairly reasonable and the quality is acceptable.

Local Rabbitry Suppliers

Many small local suppliers also sell cages. These folks are often rabbit enthusiasts who build and sell cages and other supplies as a part-time business. Local suppliers are a good source to consider, but it may be a challenge to find one near you. Talk with other rabbit raisers to find a supplier that serves your area.

Pet Shops

This may seem like the logical place to shop, but I encourage you to look elsewhere. Pet-store prices are often very high. You can find better quality and more variety at businesses, like those listed above, that specialize in rabbitry equipment.

Features of Rabbit Cages

Once you have found some sources, your next challenge will be to find the cage that is right for you and your rabbit. To do that, you'll have to make some decisions about many different types of cages.

Types of Wire in All-wire Cages

The all-wire hutch is the first choice of an experienced rabbit raiser. As you look at cages, you will discover that there are many types of wire. The most common wire used for the sides and tops of hutches is 14-gauge wire in a 1" x 2" mesh. Wire mesh that is smaller than 1" x 2" is fine, but it will probably be more expensive.

Sometimes you may have a choice of 14-gauge wire or 16-gauge wire. The 14-gauge wire is heavier and stronger. While 16-gauge wire will work for the smaller breeds, you should always choose 14-gauge if you are raising medium or large breeds. As a general rule, a cage made from 14-gauge wire may cost a little more, but it will last longer and is usually worth the extra money. You may also need to choose between wire that is welded before or after being *galvanized.* Wire that is galvanized *after* welding is stronger and smoother for the rabbit's feet.

The floor of your cage should be made of 14-gauge welded wire that is ½" x 1". This smaller mesh gives more support to the rabbit's feet, but it still has spaces big enough to allow the manure to pass through. Never use 1" x 1" or 1" x 2" wire for floor wire. Rabbits can get their feet caught in the openings and break or dislocate their rear legs.

Poor Cage Materials

Do not choose a hutch that is made with poultry wire or with hardware cloth. These products are not sturdy enough for a safe, long-lasting hutch.

Do not buy a cage if it is made out of mesh that is larger than 1" x 2". A larger mesh may be cheaper, but it is not a good idea because baby bunnies can squeeze through.

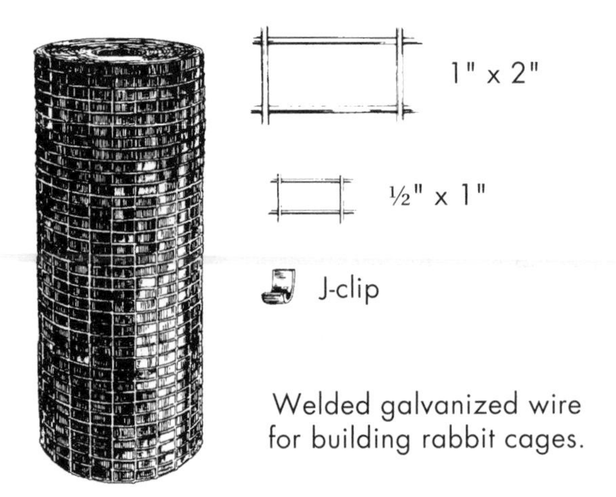

Welded galvanized wire for building rabbit cages.

Styles of Wire Cages

There are two basic styles of wire hutches. The first type is intended to be suspended so the manure falls to the ground. The second type includes a pull-out tray that the manure falls into. These cages, too, can be hung, but they are also designed to stack on top of each other. Stacking can make it a little harder to reach each cage, but it saves space and is a good feature if you have a very small indoor rabbitry. You must have very good ventilation if you plan to keep a number of rabbits indoors.

Types of Doors

The door opening is cut into the front of the cage. The door itself should be larger than the opening so that it overlaps the opening on all sides. Doors should be placed toward one side of the front of the cage, leaving enough space for the feeder and waterer.

Some doors are attached at the top and swing into the cage when opened. The nice thing about this style of door is that the door will fall back to a closed position even if you forget to latch it, and the rabbit cannot push it open. You will also see cages with doors that open toward the outside. This style makes it easier to reach into the cage, but if you forget to latch the door, your rabbit may push it open and tumble out of the cage.

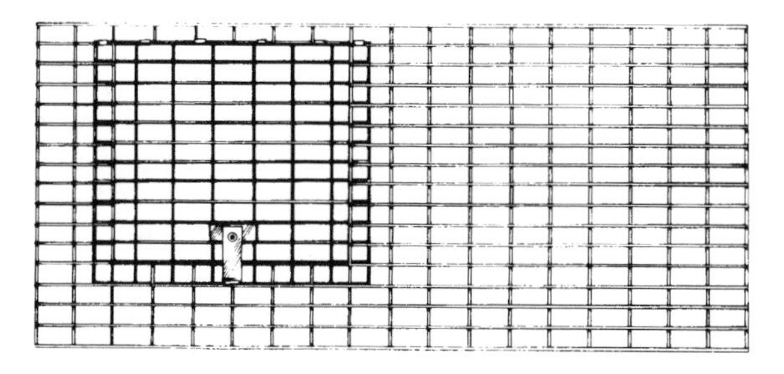

Door overlaps opening on three sides and swings into the cage.

Check the door and door opening for sharp edges. You don't want a cage that will scratch you or your rabbit, or tear your clothing. Some cage doors have a metal or plastic covering to protect you from sharp edges.

Urine Guards

A urine guard is a common extra feature on rabbit cages. Four-inch high metal strips attached around the inside base of the cage help direct the rabbit's urine so

Baby-Saver Wire

Some manufacturers offer cages made from *baby-saver wire,* a special wire mesh that can be used for cage sides. The upper portion of baby-saver wire has 1" x 2" mesh. The bottom 4 inches has ½" x 1" mesh. This smaller mesh next to the cage floor keeps babies safely inside the hutch. Baby-saver wire is more expensive than 1" x 2" wire, so cages made with it will cost more. However, if your cage will house a producing doe, baby-saver wire is a wise investment.

that it falls directly below the cage floor. A urine guard also helps prevent baby bunnies from becoming stuck in or falling through the sides of the cage. Urine guards are *not,* however, a substitute for baby-saver wire. Cages with urine guards are more expensive than those without them.

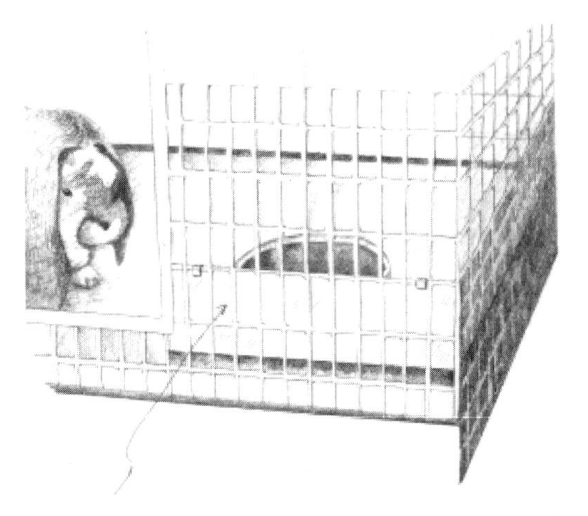

Urine guards help keep the area around the cage clean.

Feeders

Most catalog pictures show a rabbit cage with an attached feeder. However, the list price of the cage usually does not include the feeder. Be sure to read the catalog description carefully, or ask the supplier if the feeder is included. If not, you'll need to provide a feeder for the cage. (See pages 50–53 for information about feeding equipment.)

Why Build Your Own Cages?

With the help of an adult, you can learn to build your own rabbit cages. If you are planning a small rabbitry, buying cages may actually be less expensive than building them. The prices suppliers charge for cages are based on the cost of materials used to build the cages. By purchasing materials in large quantities, suppliers get a discount, and this means they can charge less per cage. When you purchase materials to build just one or two cages, you may find that your cost per cage is as much or possibly even higher than the price you would pay to buy a cage. Then, why build your own cages? There are several very good reasons:

- You can design your own cages exactly as you choose. You can make cages that best fit your space and that best meet the needs of your rabbits.
- You and your family can work together and feel good about your accomplishment.
- Once you learn how, you can build cages for others and earn money to help with your rabbitry expenses.

There are many different variations of rabbit hutches. The two that are included in this book were chosen because they are fairly simple and they can be adapted to fit into a wide variety of small rabbitries.

Building a Single Hutch

This hutch will be 3 feet long, 2½ feet wide, and 18 inches high, an adequate size for most breeds, except the giant breeds.

Materials

- ❑ 12½ feet of 18" high 14-gauge baby-saver wire or 1" x 2" 14-gauge welded wire (sides and door)
- ❑ 3 feet of 30" wide 14-gauge ½" x 1" welded wire (floor)
- ❑ 3 feet of 30" wide 1" x 2" 14-gauge welded wire (top)

More About Rabbit Housing

Additional sources for hutch plans and other information about rabbit housing are included in the following:

- Most 4-H rabbit manuals
- A.R.B.A. publications
- Garden Way Publishing's Country Wisdom Bulletin, *Build Rabbit Housing*

- ❑ J-clips (about ¼ pound)
- ❑ latch for the cage door

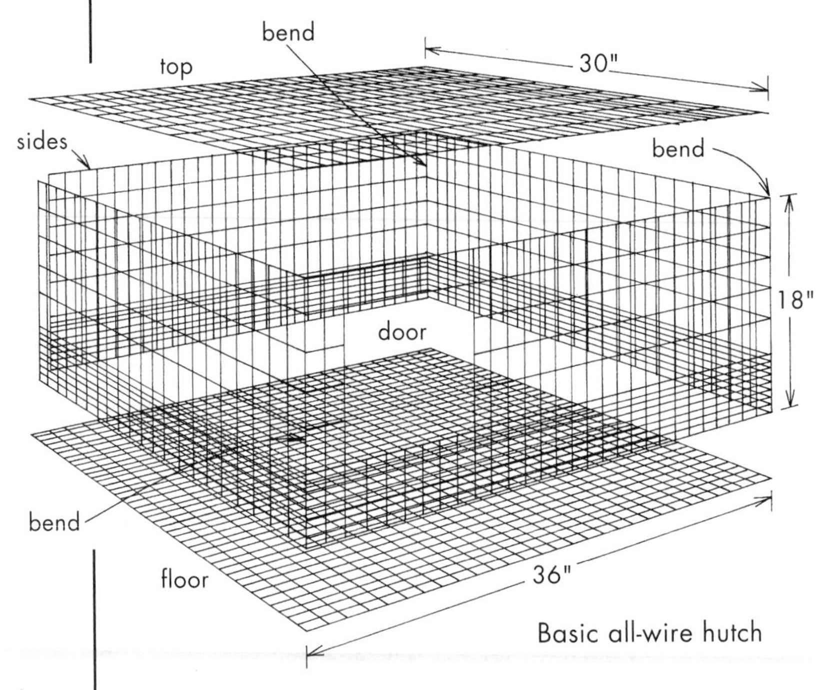

Basic all-wire hutch

Tools

To construct an all-wire cage, you will need the following tools:

- tape measure
- wire cutters
- pliers
- hammer
- J-clips and J-clip pliers
- section of 2" x 4" lumber, about 2 feet long

Sides of the Cage

1. Cut an 11-foot length of the 18" baby-saver wire.
2. Measure a distance 36 inches from one end. At this spot use the section of 2x4 lumber and bend the wire over it as shown below to form the first corner of the cage.

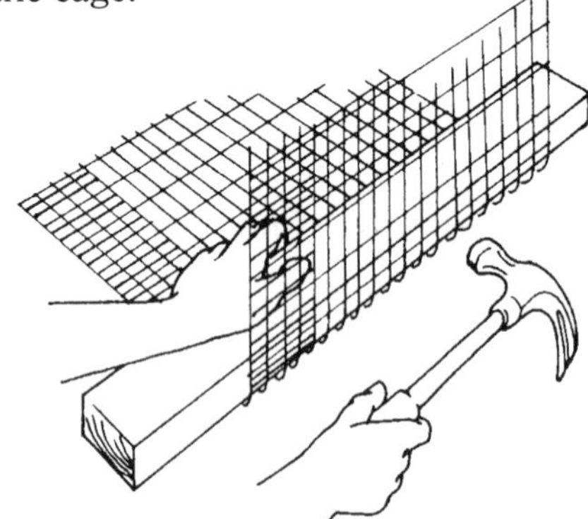

3. Measure 30 inches from the first corner and form a second corner using the same method.

4. Measure 36 inches from the second corner and make the third and final corner.

5. You will now have a rectangular shape, 30" x 36". Close the rectangle by joining the ends together with the J-clips. Clips should be placed at the very top, at the bottom, and every 2 to 3 inches in between. The four sides of your cage are now complete.

Floor and Top

6. Now you can use J-clips to attach the ½" x 1" wire to the sides of the cage, creating the floor. Remember that the smaller mesh of the baby-saver wire forms the lower part of the cage, near the floor, and the larger 1" x 2" mesh forms the upper part, near the top. The ½" x 1" wire is used for the floor because the small mesh will provide a comfortable surface for your rabbit's feet. The 1" side of the mesh should face down toward the ground; the ½" side provides a smoother surface for your rabbit to stand on and helps prevent sore hocks (see pages 70–71).

7. Use J-clips to attach the 1" x 2" wire section to form the cage top.

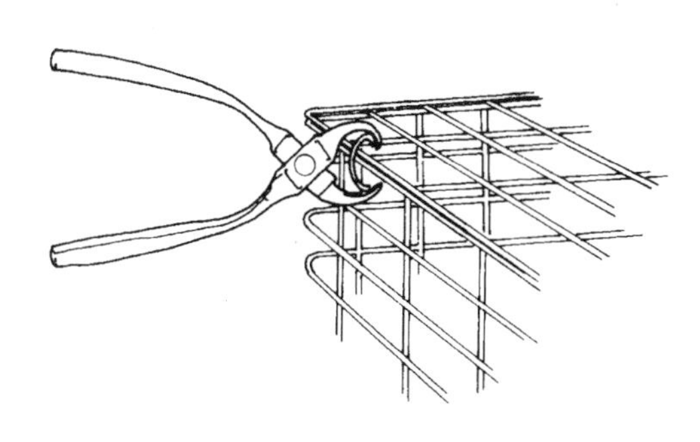

TOOLS NEEDED FOR BUILDING CAGES

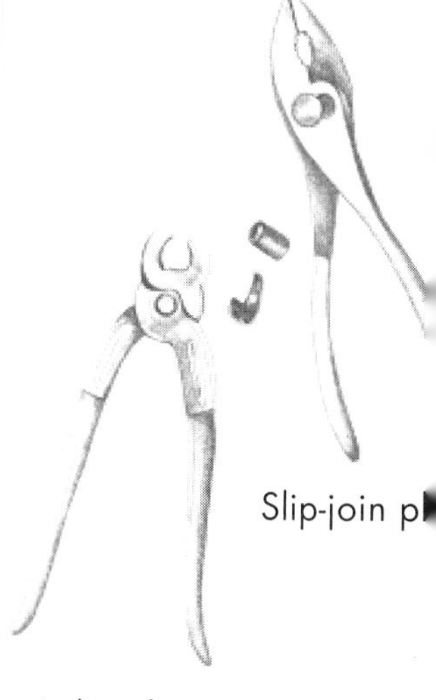

Slip-join p

J-clip pliers

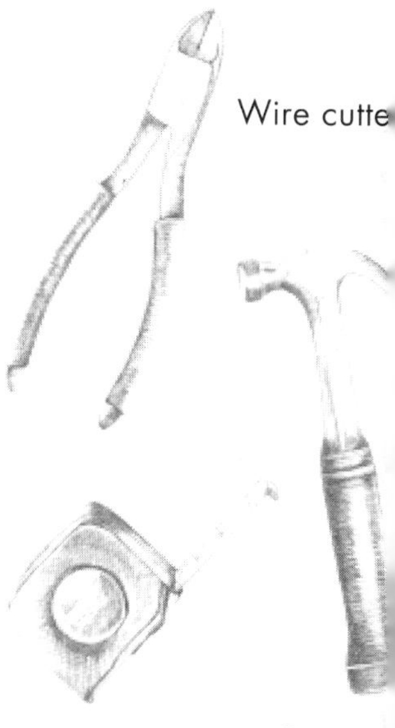

Wire cutte

Tape measure

Carpen hamm

The Door Opening

8. Before you cut the door opening, you have to make two decisions: how large the door will be and where to position it on the front of the cage. Remember, the door must be big enough to allow you to put a nest box into the cage. For most breeds, a door opening of 12" x 12" is about right. If you plan to attach an outside metal feeder, place the door toward one side of the front of the cage. The bottom edge of the door opening should be 4 inches above the floor of the cage, to prevent bunnies from falling out when the cage door is open.

9. Measure the location and size of the opening, then use wire cutters to make the opening. Make each cut so that you leave an end of wire about ¾" long sticking out.

10. After the door section is cut out, use regular pliers to bend these protruding ends over. This process takes a little more time, but it creates a smooth finish that will not scratch you or tear your clothes when you reach into the cage. An alternative would be to purchase plastic or metal door edge protectors, but this is an unnecessary added expense.

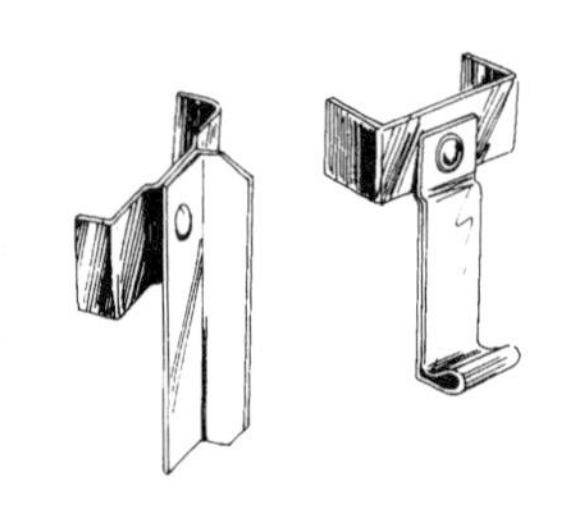

Two kinds of galvanized metal door latches

The Door

11. Make your cage door from the piece of wire remaining from the piece you cut for the sides. The door should be larger than the door opening. If the opening is 12" x 12", you will want the door to be at least 14" x 14".

12. Instead of attaching the door section to the very edge of the door opening, attach it at least one space beyond the edge. This makes it sturdier. If

you've decided to have the door open into the cage, attach it with J-clips along the top of the opening. (See page 33.) Doors that open out are generally attached along one side. Once a door latch is attached, you have completed your first all-wire hutch!

Building a Multiple-Unit Hutch

Multiple-unit hutches for rabbits are like apartment houses for people. Because adjoining units share walls, you need less wire, and the construction cost per unit is less than the cost to construct individual units. If you are planning to have several rabbits, a multiple-unit hutch may be right for you. You will use the same tools, types of wire, and building skills that are required to build a single unit.

The all-wire hutch described below will measure 7 feet in length, 30 inches in width, and 18 inches in height. It can be subdivided into four units for small

Materials for Multiple-Hutch Unit

Size	Wire for Sides (18", baby-saver)	Wire for Top and Doors (1" x 2", 14-gauge, 30" wide)	Wire for Floor and Partitions (1" x ½", 14-gauge, 30")	Number of Door Latches
2-Unit	19 ft.	8½ ft.	8½ ft.	2
3-Unit	19 ft.	10 ft.	10½ ft.	3
4-Unit	19 ft.	10 ft.	12 ft.	4

breeds, three units for medium-size breeds, or two units for large breeds.

Sides

1. Use the 19-foot length of baby-saver wire to form the walls.
2. Measure 7 feet, and use a piece of 2" x 4" and a hammer to form the first corner.
3. From this bend measure 30 inches and make a second corner.
4. Measure 7 feet and form the last corner.
5. Join the two ends with J-clips to form a large rectangle (30" x 7').

Floor

6. Cut a 7-foot length of ½" x 1" mesh. Use J-clips to connect this floor piece to the four walls. Attach the flooring wire as described in Step 6 on page 37.

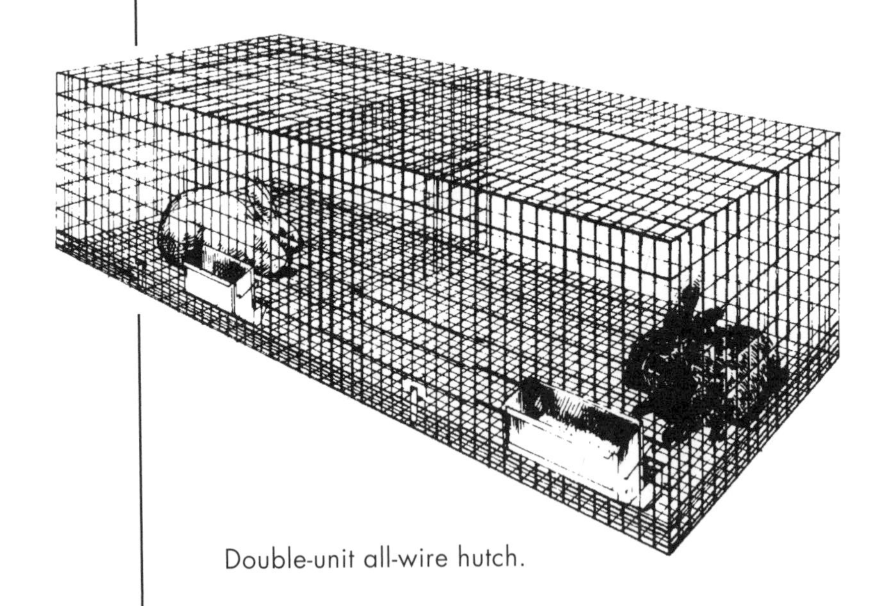

Double-unit all-wire hutch.

Interior Partitions

7. Cut the remaining ½" x 1" mesh into 18" x 30" pieces that will be used to subdivide the cage. You will need one 18" x 30" partition for a two-unit cage; two 18" x 30" pieces for a three-unit cage; and three 18" x 30" pieces for a four-unit cage. (You can also purchase solid metal or plastic dividers from rabbitry suppliers.)
8. Use J-clips to attach the partitions securely to the floor, the front, and the back of the cage. Space the partitions 21 inches apart for a four-unit cage, 28 inches apart for a three-unit cage, or 42 inches apart for a two-unit cage.

Top

9. Cut a 7-foot section of 30" wide, 1" x 2" mesh for the top of the cage. Attach the cage top with J-clips to the front, back, ends, and interior partitions.

Door Openings and Doors

10. Review Steps 8-12 on door openings and doors on pages 38–39. Cut a door opening in the front of each unit. Use the remaining 1" x 2" mesh to cut the doors. Attach the doors, and attach a latch to each door.

Frame for a Single Hutch

If your all-wire hutch will be used outdoors, you will need to adapt it so that your rabbit is protected under various weather conditions. The most common way to provide protection from the weather is to build a wooden framework that your cage sets into. Constructing this framework will be more of a challenge than cage building. You will definitely need the help of an adult who has some basic carpentry skills. Plans for a single hutch frame are shown in the illustration on page 44. You'll need to modify them for a multiple-unit hutch.

Materials

- ❑ Four lengths of 2" x 4" x 10' lumber
- ❑ One length of 2" x 4" x 8' lumber
- ❑ One 4' x 5' sheet of exterior-grade plywood (¾" thickness)
- ❑ 12d common nails (galvanized)
- ❑ 6d common nails (galvanized)
- ❑ 8 L-brackets
- ❑ Wood screws

1. Cut the two tall front legs to 60" from one of the 10-foot 2x4s.
2. Cut the two shorter back legs to 56" from a second 10-foot 2x4.
3. Use the remaining 10-foot 2x4 to make the cage support frame and the front and back roof supports. First, cut one piece from each 2x4 to 35¼". Then, cut one piece from each 2x4 to 38". These four pieces will make the cage support frame. Finally, cut the two remaining scraps to 44⅛" to make the front and back roof supports.
4. Cut the 8-foot 2x4 in half. These pieces will be the side roof supports, and will be re-cut to length later.
5. Take the four cage-support frame pieces and lay them on edge on a flat surface. They should form a

big rectangle with the 38-inch front and back pieces *between* the 35¼" side pieces. Drive two nails through the outside face of a side piece into the end of the back piece to make the corner. Nail the other three corners together in the same fashion to complete the cage-support frame.

6. Screw the L-brackets to the inside face of the cage-support frame, as shown in the illustration on page 44. Make sure the bottom of each bracket is flush with the bottom edge of the frame.

7. Mark one end of each of the legs with a pencil to indicate the bottom, and to avoid confusion during assembly. Measure up from the bottom of each leg 32 inches, and draw a square line across the inside face of each leg at this point. This mark represents the height of the cage-support frame.

8. Nail the legs to the ends of the frame, as shown, making sure the bottom of the frame is even with the lines marked on the legs.

9. Nail the front and back roof supports to the legs. The top edge of the front roof supports should be slightly higher than the top of the front legs. This will prevent the front legs from being in the way when it comes time to put on the plywood roof.

10. Take one of the side roof supports and hold it in place against the legs on the right side of the hutch. Mark the angled cut with a pencil, cut the piece to size, and nail it to the front and back roof supports. For added strength, nail it to the tops of the legs as well. Repeat this procedure for the left side.

11. Use scraps of plywood to cut four isosceles triangles, with the equal sides being 12" in length. Nail these braces to the hutch as shown with 6d nails.

12. Nail the 4' x 4' plywood roof in place with 6d nails. The roof should overhang all four sides, to help

keep rain or snow out of the hutch. Plan a larger overhang along the front to give extra protection to the attached feeders.

Advantages of a Wooden Frame

This type of frame has features that make it adaptable to a variety of situations:

- Sides can be left open for good air circulation.
- Removable, solid wooden sides or plastic sheeting can be attached for cold weather protection.
- The cage can be removed from the frame, so cleaning is easy.
- Since the cage can be removed from the frame, it is easy to move an expectant doe to a warmer location for a winter kindling.

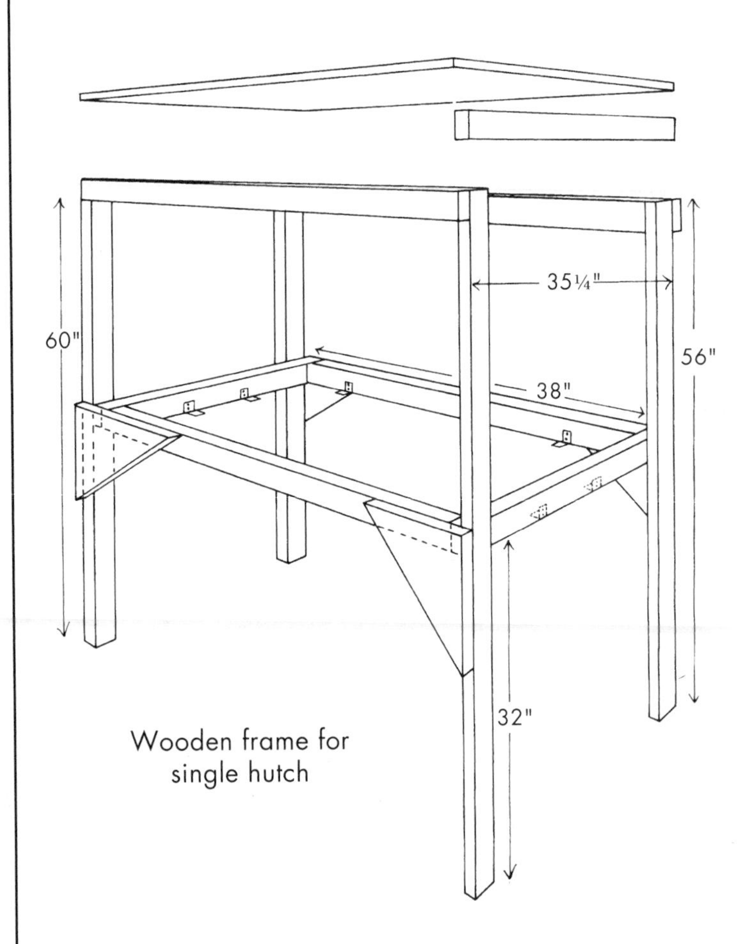

Wooden frame for single hutch

Home Away from Home: Rabbit Carriers

If you plan to show your rabbit or need to move it from one place to another for any reason, you'll need a rabbit carrier. Like cages, the best rabbit carriers are made of wire mesh. While you can purchase a good rabbit carrier, you can also build one yourself. If you have built a cage, then building your own carrier will be easy and economical. Many small pieces left from cage construction can be used to build a carrier. If you haven't built a cage and don't plan to, and if you would have to purchase wire cutters, J-clips, J-clip pliers, and wire to make a carrier, then making your own carrier is probably not worth the effort or expense.

Building carriers can be a good group project. Working with others is fun, plus you can share tools and save money by buying supplies in larger quantities.

Bottom Tray

One of the largest costs involved in building your own carriers is the cost of the metal pan that encloses the bottom. You can cut this cost by using a substitute such as a plastic cat litter tray or a large roasting pan found at a local garage sale. One of my resourceful 4-H fathers uses the plastic containers in which fresh fish are shipped. The fish department at your local supermarket will usually give these away.

Tips on Building a Carrier

The sides of the carrier can be made from one long piece of wire mesh or from four smaller pieces. The length of the four sides will vary, based on the size of the bottom tray. The height of the carrier sides will vary according to the size of the breed you are building the carrier for. Small breeds require a height of 8 inches, medium breeds require 10 inches, large breeds require 12 inches, and giant breeds require 14 to 16 inches.

Before cutting wire for the sides, remember to add an extra 2 inches to the height. This extra height allows you to attach the floor 2 inches above the tray so droppings can pass through. The floor of your carrier should be made from ½" x 1" welded wire. (Be sure to place the ½-inch side toward the top to protect the rabbit's feet.)

Rabbit carriers are designed so that the top opens to make it easy to reach in and safely remove the rabbit. Plan to allow an extra 1 to 2 inches on each end and on one side of your top piece. This extra wire can be bent so that it overlaps three of the sides. The straight edge is attached with J-clips to form a hinge on the fourth side, allowing the top to open and close. Attach a door latch or hook so you will be able to securely close your carrier. Recycling the clip from an old pet leash will also make a secure closure.

S-hooks and springs are commonly used to attach the cage to the tray. These items are fairly inexpensive, and they make it very easy to remove the tray for cleaning.

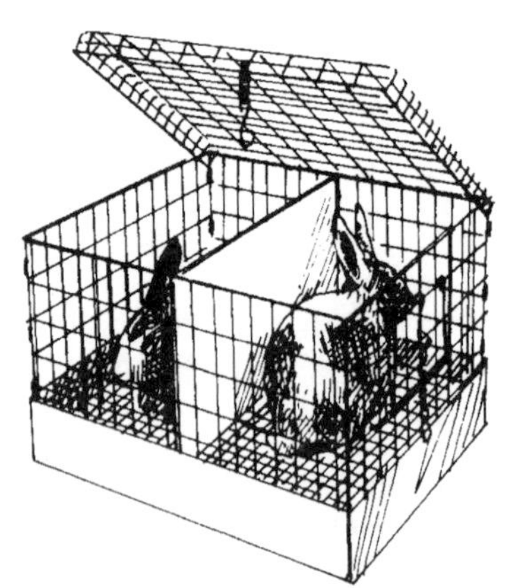

Rabbit carrying cages

Feeding Your Rabbit

CHAPTER 5

Providing proper feed for your rabbits is a very important part of successful rabbit raising. A healthy, balanced diet is based on providing commercial rabbit pellets and fresh water. In the past, rabbit owners had to mix different feeds in order to get the proper nutritional balance for their animals. Nowadays, stores carry feed that is already balanced. Even though this simplifies your job, a good feeding program is much more than simply filling up the feed dish.

What to Feed

Commercial rabbit pellets are the best and easiest way to provide proper nutrition for your rabbits. A great deal of research goes into producing a good, balanced ration, and feed companies are better qualified than you, the breeder, to do this.

Commercial feeds do vary, and different feeds are available in different locations. Talk with other breeders and with your feed store clerk to learn about brands of feed that are available in your area.

You'll also learn that even under the same brand name there are often several different types of feed

PAMELA T. BERNARDINI

Providing proper feed is an important part of raising rabbits.

available. These types of feed are planned to meet the needs of different rabbits at different stages in their lives. The feed ingredient that usually varies is the protein content. Mature animals usually need feeds that provide 15 to 16 percent protein. Young growing animals and active breeding does are often fed a ration with a higher level of protein, 17 to 18 percent.

Understanding Feed Labels

The U.S. government requires that all rabbit feed contain at least a certain amount of protein and fat. A feed may contain more than those amounts, but not less. The amount of protein and fat is listed on the feed package. The package also lists the fiber content. A package label may say that the feed has "no more than 18 or 20 percent fiber," but this tells you only the *maximum* amount of fiber in the feed. You need also to know the *minimum* amount it contains. It is very important that rabbits eat feed containing at least 16 percent fiber. If they get less than that, they may have problems with diarrhea.

Water

No matter which type of feed you provide, your rabbits will not eat well unless you provide the most important nutrient of all—*water!* We often do not think of water as food, but it is essential to the health of your rabbits. Rabbits simply do not thrive without a constant supply of clean water.

Salt

Salt is another ingredient that is important to a balanced feed. It is not necessary to provide a salt spool for your rabbit (and they cause your cage to rust, as well). You may see advertisements for inexpensive salt spools that can be hung in each cage, but the commercial pellets you use as feed already include enough salt to meet your rabbit's needs.

Hay

Many rabbit raisers also feed hay to their rabbits. Rabbits enjoy good-quality hay, and hay adds extra fiber to their diet. If you feed hay, be sure it smells good, is not dusty, and is not moldy. Feeding hay will add more time to your feeding and cleaning chores.

If you choose to feed hay on a regular basis, you will want to have hay mangers in each cage. Hay mangers are easy to construct from scraps of 1" x 2" wire (if you make your own cage, you can use the leftover wire). If your cages are inside and are of all-wire construction, you can also just place a handful of hay on the top and the rabbits will reach up and pull it through the wire. Try to avoid setting hay inside on the floor of the cage, because it will soon become soiled and droppings will collect on it. This can cause disease and parasite problems.

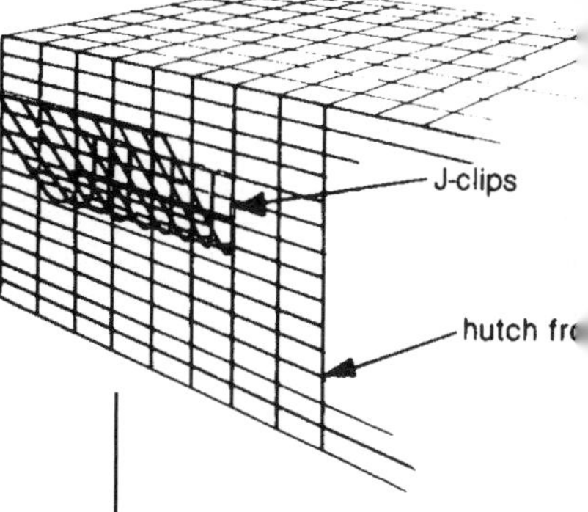

To make a simple hay rack, bend a scrap of wire and attach it to the side of your cage with J-clips.

What About Special Feeds and Treats?

Some breeders give their rabbits a long list of other feeds. Experienced rabbit raisers, especially those trying to get their animals ready for shows, have special feeds that they feel give their animals an edge. Extras such as oats, sweet feeds, corn, and sunflower seeds are a few of the special feeds they provide along with pelleted feeds. However, any additions to the diet can cause upsets to the rabbit's digestion. It is best to keep to the simple diet of pellets and water, and maybe some hay, and leave the additional feeds until you have experienced several years of successful rabbit raising.

Many new rabbit raisers think that they are being nice to their animals if they give them fresh greens and vegetables as "treats." In fact, lots of these can actually be harmful to your rabbits. Young rabbits are especially sensitive to too many treats, so it is best not to give them any. As the rabbit matures, its digestive system can handle more things, but treats should never become a large part of your rabbit's diet. Some common treats that are okay for mature rabbits are slices of apple or a piece of carrot. Lettuce and cabbage are *not* good treats for rabbits of any age, because it gives them diarrhea.

Feeding Equipment

Once you know what to feed your rabbits, you need to find a way to make this feed available to them. Pellets are usually fed from crocks (heavy plastic or pottery dishes) that sit on the cage floor, or from self-feeders that attach to the side of the cage. If you choose to use crocks, select one that is heavy enough so that your rabbit cannot tip it over and waste its feed. To make feeding easier, place your crocks so they are close to the door of the cage. Never place the crock in the area

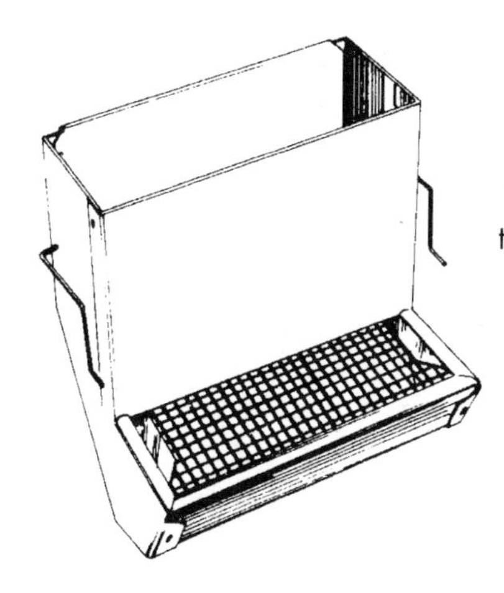

This self-feeder clips to the cage wire and can be filled from outside the cage.

that your rabbit has chosen as its toilet area. Because rabbits can soil the food in crocks, feed only what will be eaten between each feeding time.

Self-feeders attach to the wall of the cage. This type of feeder is filled from outside the cage, thus making it easier for you to care for your rabbit. If you use a self-feeder, choose the type that has a screened, not solid, bottom. The screen allows the *fines* (small, dusty pieces that break off from the pellets) to pass through. Rabbits will not eat dust from feed, so eliminating the build-up of fines is a plus that self-feeders offer. Another advantage of self-feeders is that young rabbits are less likely to climb into them and dirty the feed.

Make Your Own Feeder

Many new rabbit raisers start out by using containers that they find around the house. Although small bowls are the right size, they tip easily and you soon find you are spending more money to replace the wasted food than new feeders would have cost.

If you need to keep costs low as you start out (and most everyone does), make your own simple feeder from a large (12-ounce) tuna fish can. Tack the can to a 10-inch square of ½-inch plywood. The plywood base

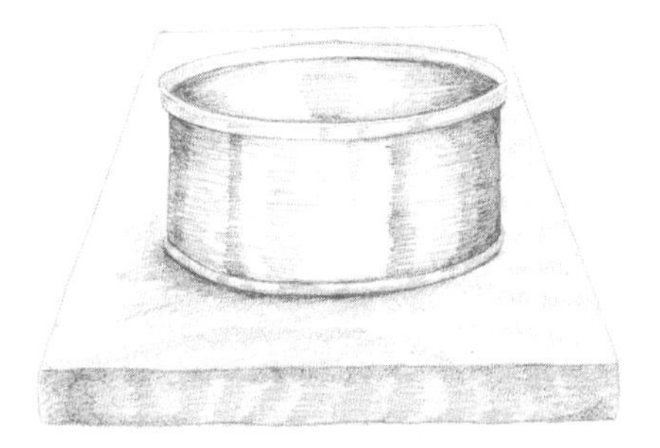

Homemade, tuna-fish-can feeder.

Crocks can be used for water as well as feed.

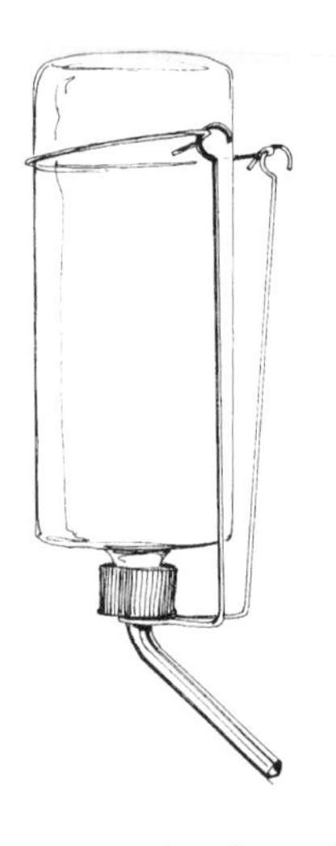

Plastic water bottle with metal tube

will make your feeder hard to tip, but you will still be able to lift it out of the cage for regular cleanings.

Watering Equipment

As with feed, there are several ways to provide water to your rabbits. Crocks can be used for water as well as for feed. If you live in an area where freezing temperatures are common and you keep your rabbit outside, you might try heavy plastic crocks. You can remove frozen water from these crocks without breaking them or waiting for them to defrost. Hit plastic crocks against a fence post or other solid object to dislodge the frozen water. This can save a lot of time during winter feedings.

Another type of waterer is a plastic water bottle, which hangs on the outside of the cage. It has a metal tube that delivers water to the rabbit. This type of waterer keeps water cleaner than crocks do, but waterers freeze easily and therefore cannot be used in cold weather.

Homemade Waterers

Needless to say, you can't nail a can to a board and use it as a water container—the water would run right out of the nail holes. Instead, make a homemade weighted dish from empty bleach bottles. For each waterer, you need a 1-gallon bottle and a ½-gallon bottle. Cut around each bottle about 4 inches from the bottom. Place the ½-gallon piece in the center of the gallon piece. Fill the space between the two with cement. When the cement hardens, you will have a low-cost, hard-to-tip waterer.

Another type of homemade water container can be made from a plastic bleach or soda bottle and a drinking valve, such as that used in automatic watering systems. If your local farm store does not have these valves, you can order one from a rabbitry supply company (see listing on page 141). Ask an adult to help you cut or drill a hole into the side of the bottle near the bottom. Spread an epoxy cement on the rear part of the valve, and then insert it into the hole you made. Use wire to hang the finished waterer on the outside of the cage. The good thing about this type of waterer is that your rabbit cannot dirty the water. Like other hanging water bottles, however, it freezes easily.

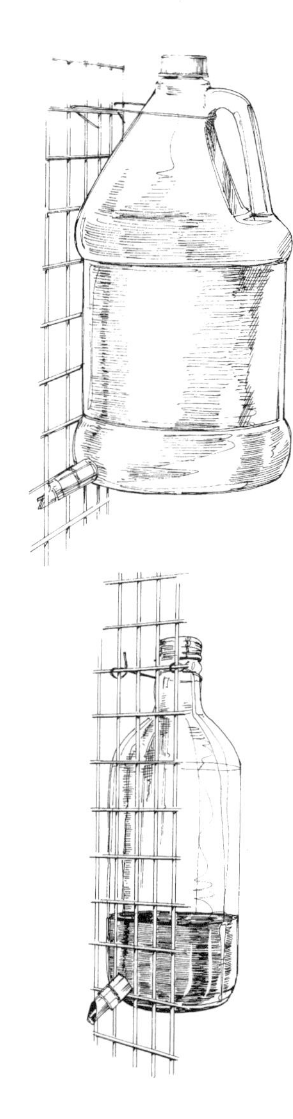

Homemade water bottles fitted with drinking valves (top, bleach bottle; bottom, soda bottle)

When to Feed

If we study the feeding habits of wild rabbits, we can apply what we learn to feeding our domestic rabbits. Are there wild rabbits that live near you? Have you ever noticed that you don't see them very often during the day, but they often appear in the early morning or late afternoon? In nature, rabbits have learned that it is safer to stay in a safe place during the day. In late afternoon, hungry wild rabbits come out of hiding. By this time, they are ready for their big meal of the day. Wild rabbits are *nocturnal animals,* active throughout the night and into the early morning hours.

If our tame rabbits could pick a mealtime, they would probably agree with their wild relatives and choose late afternoon. Using this information, most rabbit breeders feed the largest portion of their rabbits'

Nocturnal. *An animal that is active at night.*

daily ration in the late afternoon or early evening.

How often should your rabbits be fed? Although different breeders develop a schedule that works for them, two feedings a day—one in the morning and one in the late afternoon—seem to be most common. If you choose to feed twice a day, serve a lighter feeding in the morning and a larger portion in the afternoon.

Feeding Time Is Check-Up Time

If you have self-feeders and water bottles, you may feel comfortable with feeding only once a day. Feeding time, however, is important for more than just feeding. This time spent with your rabbits is an opportunity for observing many details that are important. This is the time to notice situations such as

- Has a youngster fallen out of the nest box?
- Is an animal sick or not eating well?
- Does a cage need a simple repair before it becomes a major repair?
- Does your rabbit need extra water due to extreme heat or freezing conditions? Remember that the overall health of your rabbit will be better if clean, fresh water is available at all times.

Because a lot can happen to your rabbits in a 24-hour period, I strongly recommend that you feed, water, and observe your animals at least twice a day. The serious rabbit breeder will see this as a worthwhile investment in time.

Whatever feeding schedule you choose, remember that it is very important to follow that schedule every day. Your rabbits depend upon you, and they will settle into the schedule that you set. Avoid variations in feeding times. If you regularly feed at 7 A.M., you should do so even if it's Saturday and you want to stay in bed a little longer. Plan ahead for times when you cannot be there to feed your rabbits. Be sure other family members are familiar with your feeding program so they are able to help out from time to time. If your family includes more than one rabbit raiser, you can take turns with feeding responsibilities. You can also trade off other chores, such as washing dishes, in exchange for filling in when you have to be away.

How Much to Feed

Knowing how much to feed is probably the most complicated part of feeding your rabbits. There is no set amount that is right for all animals, but the following chart will give you some average amounts.

In order to use the guidelines in the chart, you need

Feeding Chart

	Amount of Pellets Needed Each Day		
	Small Breeds	**Medium Breeds**	**Large Breeds**
Bucks	2 oz.	3–6 oz.	4–9 oz.
Does	3 oz.	6 oz.	9 oz.
Doe, bred 1–15 days	3 oz.	6 oz.	9 oz.
Doe, bred 16–30 days	3½–4 oz.	7–8 oz.	10–11 oz.
Doe, plus litter (6–8 young), 1 week old	8 oz.	10 oz.	12 oz.
Doe, plus litter (6–8 young), 1 month old	14 oz.	18 oz.	24 oz.
Doe, plus litter (6–8 young), 6–8 weeks old	22 oz.	28 oz.	36 oz.
Young rabbit (weaned) 2 months old	2–3 oz.	3–6 oz.	6–9 oz.

to have a simple way to measure how much you feed to each animal. You can make your own measuring container from empty cans. For example, a 6-ounce tuna fish can filled level to the top holds 4 ounces of pellets. The same can filled so that the pellets form a small mound above the lid gives a 5-ounce serving. You can use a larger can and mark several different levels to indicate the amounts you most commonly feed.

How to Make a Feed Measure

Materials

- ❑ Empty cans
- ❑ Permanent marker
- ❑ Scale that indicates weight in ounces
- ❑ Pellets

Lots of adults who are on diets have small scales that they use to weigh the food they eat. These scales measure in ounces and will work well for this project. See if you can borrow a diet scale. You need to adjust for the weight of the can, so place your empty can on the scale, and set the weight dial back to zero. If the scale that you use does not adjust for the weight of the can, you will have to remember to figure this into your measurements. For example, if your can weighs 1 ounce, to find a 4-ounce level, add pellets until the scale reads 5 ounces (1 ounce for the weight of the can, plus 4 ounces of pellets).

1. Add pellets to the empty can until the scale reads the number of ounces you want to feed.
2. Mark a line at that level on the outside of the can. You may also want to mark the number of ounces.

3. Add or subtract pellets to indicate other feed levels. If you are feeding different rabbits different amounts, mark which of your animals receives each amount.

4. It might be helpful to mark each level on the inside of the can also.

Feed guidelines are exactly what their name says, *guidelines only.* As you get to know your rabbits

Homemade feed measure

better, you will find that some need a little more feed and some need less. A simple way to judge who needs more or less is to feel your rabbits regularly. Just a few pats can give you an idea of their condition. If you stroke each rabbit from the base of its head and follow along the backbone, you will soon learn to tell who is too fat and who is too thin. When you run your hand over the rabbit, you will feel the backbone below the fur and muscles. As you feel the bumps of the individual bones that make up the backbone, they should feel rounded. If the bumps feel sharp or pointed, your rabbit can use an increase in feed. (Note: Extreme thinness may also indicate health problems. See Chapter 6.) If you don't feel the individual bumps, it probably means there is too much fat covering them, so that rabbit will do better on less feed. If you need to adjust the amount you feed, make the change gradually, not all at once. Unless the rabbit is extremely thin or extremely fat, increase or decrease by about 1 ounce

Some Reasons a Rabbit's Feeding May Need Adjusting

- A young, growing rabbit needs more feed than a mature rabbit of the same size.
- Rabbits need more feed in cold weather than in hot.
- A doe that is producing milk for her litter needs more food than a doe without a litter.

Stroke your rabbit along its backbone regularly to judge whether it is getting the proper amount of food.

each day. A small plastic scoop that is used to measure ground coffee holds about 1 ounce of pellets. Use this scoop to add or subtract feed, 1 ounce at a time. It will take awhile for the change to show on your rabbit, but you should see and feel a difference in 1 to 2 weeks. It's good to form the habit of giving each rabbit a weekly check. It takes only a few seconds to feel down the back of an animal. If you check regularly, you will be able to adjust feed before your rabbit becomes extremely fat or thin.

In general, overfeeding is a more common problem than underfeeding. For one thing, overfeeding means it costs you more to care for your rabbits. In addition, if you are breeding your rabbits, you'll find that overweight rabbits are generally less productive than animals of the proper weight. In fact, fat rabbits often do not breed, and an overweight doe that does become pregnant can often develop problems.

Taking Care of Your Rabbit's Health

We all know how it feels to be sick—it is not fun. We want to do all we can to keep our rabbits from feeling sick. How can you keep your rabbits healthy? You can provide H.E.L.P.

H is for proper housing. Hutches should offer protection from extremes in temperature, from winds, and from rain and snow. It is very important that hutches are cleaned often. Germs that cause diseases like to live in dirty places. A clean hutch doesn't give germs a place to grow.

E is for the rabbit's environment. Do you protect your rabbit from loud noises? Do you keep scary animals away from your rabbit? Do you provide shade on hot days? If you do, these are some ways you can make your rabbit's environment more comfortable. By removing things that cause stress or upset to your rabbit, you can actually contribute to its health.

L is for looking. One of the most important things you can do each day for your rabbit is to look at it, or observe it. This is the way you get to know what is normal for your rabbit, and you are then likely to notice when it is acting differently. Changes in how your rabbit acts or looks can often indicate that it is not feeling well. Use your other senses, too. A strong smell of ammonia indicates poor ventilation or unclean conditions. The sooner a problem is spotted, the better your chances of solving the problem before it becomes too big to solve.

P is for proper nutrition. Feeding the proper amounts of the right foods on a regular feeding schedule is very important. Don't forget the importance of lots of fresh, clean water! Be familiar with what your rabbit normally eats. A change in eating habits is often one of the first signs that a rabbit is not feeling well.

A Health Checklist

Occasionally, no matter how much H.E.L.P. you give to your rabbits, an animal becomes sick. If you suspect that an animal is not well, use the following checklist to help you pinpoint what might be wrong.

- ❑ yes ❑ no The rabbit is eating less than its normal amount of feed.
- ❑ yes ❑ no The rabbit's eyes look dull, not bright.
- ❑ yes ❑ no There is a discharge from its eyes or nose.
- ❑ yes ❑ no The rabbit's droppings are very soft (has diarrhea).
- ❑ yes ❑ no The rabbit's coat is very rough looking.
- ❑ yes ❑ no The rabbit is dirty and has manure stuck to its underside.
- ❑ yes ❑ no The rabbit is having trouble breathing or is breathing very rapidly.

Common Health Problems

If you checked one or more of the above items, you should identify and treat the problem as soon as possible. Here are some common rabbit ailments.

Abscesses

An abscess is the body's way of getting infection out of the body. Sometimes a cut or scratch gets infected, and an abscess forms. If you have ever had a small cut become infected, you may remember that after a while, the area swelled up and then a white substance (pus) drained out. This was an abscess.

Abscess. *A collection of pus caused by infection.*

If an abscess forms on your rabbit, you sometimes feel it before you see it. When you pat your rabbit, you may feel a bump. If you part the fur and look, you will see a raised area. If this area looks reddish and feels warm, it is probably an infection. Over time, an abscess gets larger and the skin covering the area opens to allow the built-up fluid, or pus, to escape. This fluid is usually thick and white. Although abscesses will drain by themselves, the pus carries germs that can infect other areas of the body. To stop the spread of these germs, you should be there when the abscess drains. Since you can't be sure when that will happen, treatment for an abscess includes starting the draining process. If your rabbit has an abscess, call your veterinarian. If you can't get professional help, ask a parent or friends to help with the following treatment:

Treating an Abscess

1. One person should hold the rabbit while the other uses medicinal disinfectant to clean the area around the abscess. It may be easier to do this if you clip away some of the hair from the area.

2. Use a sharp, clean instrument such as a razor blade to make a small opening to let the pus out. *Razor blades are very sharp so be careful!* Because the pus in the abscess has been pushing very hard against the rabbit's skin, this small cut will probably not hurt your rabbit. In fact, the cut releases the pressure of the pus, so it will actually take pain away from the area.
3. Use clean cotton balls to help push the pus out.
4. Use medicinal disinfectant to wash out the wound and the surrounding area.

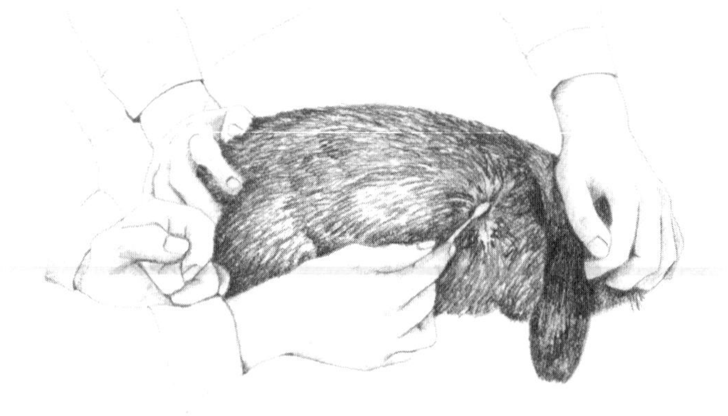

Use a cotton ball or swab to apply antibiotic cream to the wound.

5. Apply an antibiotic cream to the wound. (You can use an over-the-counter medication sold for human use.) You do not have to bandage the area. (The rabbit would only chew it off!)
6. Throw away the cotton balls or swabs in a proper place and carefully wash your hands with soap and warm water.
7. Check the wound daily and apply more antibiotic cream until it is healed.

Parasites

A *parasite* is a creature that uses a living animal as its food source. Sometimes you may notice a large bump that feels like an abscess and may even have some pus. It is not an abscess, however, if there is a hole near the top of the bump. If you look carefully, you may see a dark, wormlike shape inside the hole. This wormlike shape is actually a parasite. In this case, a fly called a *Cuterebra fly* (not a common housefly) has laid an egg on your rabbit. The egg hatched and is now growing under your rabbit's skin. In its larva stage, it looks more like a worm than a fly. If left alone, this parasite will use your rabbit as a food source to help it grow into an adult Cuterebra fly. The larva crawls out of the hole in your rabbit and develops into a flying insect.

Cuterebra larvae are harmful to your rabbit and need to be removed. Removal is usually no more difficult than treating an abscess, but it is best to seek the help of your veterinarian.

Parasite. *A creature that uses a living animal as its food source.*

Coccidiosis

A rabbit may have diarrhea because it does not have enough fiber in its diet (see page 48), or because it has a disease of the digestive system. One disease that causes diarrhea is *coccidiosis*. This is a hard word to say: cock-sid-e-O-sis. The disease it names, however, is easy for your rabbits to get. Coccidiosis does not usually kill rabbits, but if you don't treat it, it can. Coccidiosis may cause any of the following symptoms:

- Soft droppings (diarrhea)
- Fur that looks rough, not smooth
- Animal not growing as well as you think it should
- A pot belly, yet rough over the backbone

Coccidiosis is often seen in young litters of rabbits. The little organisms (called *protozoans)* that cause this disease are present in droppings and in soiled feed and bedding. A cage that was a clean home for one doe soon becomes home to Mom and eight growing youngsters. More bunnies share the same space, and so it is much more likely that manure builds up or feed dishes become dirty. Even if you make an extra effort to keep things clean, it is easy for the youngsters to pick up this disease-causing organism. Older rabbits living alone can also get coccidiosis. No matter what their age, rabbits kept in clean living conditions are much less likely to suffer from this disease.

If you need to treat your rabbits for coccidiosis, you should provide a two-part treatment, one dealing with the housing and the other, medical treatment for the animals.

PAMELA T. BERNARDINI

One of the best preventatives against coccidiosis is to keep your rabbit's cage clean.

Coccidiosis Treatment: Housing

- Remove all soiled bedding and food.
- Scrape all built-up manure from the cage at least every three days.

- Clean cage with a chlorine-bleach-and-water solution (1 part household bleach to 5 parts water). Let dry thoroughly.
- If litter is old enough (5–6 weeks), separate them into smaller groups. Of course, additional cages are necessary for this.

Coccidiosis Treatment: Medical

The medicine used to treat coccidiosis is *sulfaquinoxaline*. You should be able to find this in most farm supply stores, probably as a treatment for coccidiosis in calves, pigs, or chickens. Look for a label that lists sulfaquinoxaline as an ingredient. Although the label will probably not give directions for treating rabbits, you can use the dosage suggested for poultry or mink. If you purchase the medicine from a rabbitry supplier, it will include guidelines for rabbits.

Wash out a used milk container and label it with the name of the medicine you are using. Fill the container with water, then measure the proper amount of medicine into the water.

Provide the medicated water for 5 days, or as recommended on the label. During this time, this should be the only source of water you give to the animals being treated. If your rabbit doesn't like the taste of the treated water, add a bit of Jell-O powder to it (about 1 teaspoon to a gallon of water). One suggested treatment plan is as follows:

- Sulpha-treated water is offered for 5 days.
- Plain water is offered for 10 days.
- Sulpha-treated water is again offered for 5 days.
- Repeat this sequence two to four times a year.

This schedule is very important. During the first few days of sulpha treatment, the original organisms that made your rabbit sick are killed, and your rabbit will seem to be much better. The large population of organisms that cause coccidiosis lay eggs, however, and a second period of medication is necessary to control those eggs that hatch later.

If the animal is being raised for meat, it must not have been given medication within a certain number of days before it is sold. Follow the instructions on the label.

Colds and Snuffles

When a rabbit has a cold, it shows some of the same symptoms as you do when you have a cold—such as sneezing and having a runny nose. Unfortunately, 90 percent of the rabbits that seem to have colds actually have a much more serious disease called *Pasteurella multocid*, or *snuffles*. Although a cold can be cured, snuffles cannot. Prevention is the only way to protect your rabbits against snuffles, and the best prevention is to keep your rabbitry clean and well-ventilated.

Your rabbit's sneeze is usually the first sign of a cold or snuffles. Of course, lots of things can make a rabbit sneeze. If you hear sneezing, be sure to look for other signs of illness, too. If you see a white discharge from its nose, your rabbit probably has a cold or snuffles. Look at your rabbit's front paws. Is the fur matted and crusty? If it is, this shows that your rabbit has been wiping this runny nose. Matted front paws are another sign of a cold or snuffles.

Although rabbits and people don't give each other colds, rabbits can give colds to each other. If you suspect that a bunny has a cold or snuffles, therefore, move it away from your other rabbits. Also, be sure to disinfect its cage before moving another rabbit in.

A rabbit with a cold or snuffles is going to make other rabbits sick. If you have other rabbits, therefore, it is best to get rid of the sick rabbit by humanely putting it down. If you have only one rabbit, you can try to treat it.

Scientists are trying to find an antibiotic that will

cure snuffles in rabbits. Let's hope their work is successful.

Ear Canker

This condition is caused by tiny mites that burrow inside the rabbit's ear. A rabbit with ear mites will shake its head and scratch at its ears a lot. The inside of an unaffected rabbit's ear is nice and clean. The inside of an ear with ear mites looks dark and crusty.

Mites are more likely to appear in rabbits that live in dirty cages, so be sure to keep your cages clean.

Mites can easily be treated with common household oils, such as olive oil, cooking oil, or mineral oil. Mites breathe through pores in their skin. If they come into contact with oil, it blocks these breathing pores and they soon die.

Check all your rabbits. Ear mites can move to other animals, so other rabbits may also be infected.

Treatment for Ear Mites

- Use an eye dropper to put several drops of oil that contains a miticide, or olive, cooking, or mineral

To treat for ear mites, use a cotton swag to remove loose, crusty material in the ear.

oil into the rabbit's ear. Gently massage the base of the ear to help spread the oil around.

- Use cotton swabs to remove some of the loose, crusty material.
- Repeat this treatment daily for 3 days, wait 10 days, and then repeat the treatment for 3 more days; wait another 10 days and repeat the treatment again for 3 days.

Malocclusion

Malocclusion. *A defect in which teeth do not meet properly.*

A rabbit's teeth continue to grow, like your finger- and toenails—only faster, about ½ to ¾ inch per month. In a normal rabbit, the top front teeth slightly overlap the two bottom teeth. The normal chewing that happens when the rabbit eats keeps its teeth at a proper length. If the rabbit's teeth do not meet properly, then this normal wearing down doesn't occur and the teeth can grow very long. This condition is called *malocclusion—or buck or wolf teeth.* The top teeth may curl around and grow toward the back of the rabbit's mouth, like a ram's horns. The bottom teeth may grow out in front, instead of in back of, the top teeth. These bottom teeth may grow so long that they stick out of the rabbit's mouth, and can even grow into its nostrils. It may be hard for the rabbit to close its mouth. When the rabbit tries to chew, these long teeth can stick into it and cause it pain; they also make it hard for the rabbit to pick up its food.

If you suspect malocclusion, remove your rabbit from its cage to check its teeth carefully. This is a situation where your handling skills can really pay off, because you must turn your rabbit over to get a good look. (See pages 24–26 on handling.) To help your rabbit regain its ability to comfortably eat its food, you will have to trim these overgrown teeth.

Treatment for Malocclusion

- It is easiest to have a helper for this job—one person to hold the rabbit securely and another to do the trimming.
- Use fingernail clippers to trim the teeth as close to normal length as possible. Be careful not to cut shorter than normal, or you may cause bleeding and discomfort. You may have to trim each tooth more than once to get it down close to normal size. You may need a larger tool for some large breeds of rabbits. Your rabbit may not like having you stick the clippers in its mouth, but the actual clipping does not seem to hurt it, because there are no nerves in the teeth.
- Check your rabbit every 2 or 3 days to see how its teeth are coming along. Trimming should enable your rabbit to begin to eat normally again.

Most cases of malocclusion are inherited. This means that the rabbit was born with the condition or with the tendency to develop the condition. Rabbit breeders do not want to raise more animals with this condition, so animals with malocclusion should *never* be used as breeders. Neither should they be shown.

Some rabbits develop malocclusion from an injury. This occasionally happens when a rabbit chews on its wire cage.

Whatever the cause, once the teeth are out of position, they rarely return to the proper position. In most cases, the teeth must be clipped every 2 to 3 weeks for the rest of the rabbit's life.

Sore Hocks

The *hock* is the joint in a rabbit's hind leg between the upper and lower leg bones. The rabbit carries most of

Signs of Malocclusion

- Your rabbit is losing weight, but does not seem sick
- Your rabbit drops its food or seems to have difficulty chewing when it eat

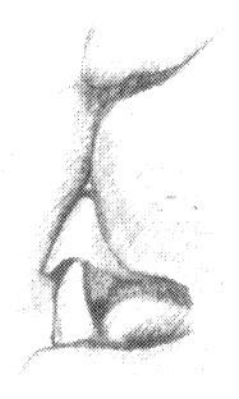

(Top) Properly aligned teeth. (Bottom) Teeth are overgrown because they are not properly aligned.

its weight on its hind feet, so the area between the hock and the foot can take a lot of wear and tear. Although the rabbit has fur on this part of the leg, it is not always thick enough to protect it. Sometimes the fur wears away and the unprotected skin develops sores. A sore hock can cause a lot of discomfort.

You can suspect sore hocks if your rabbit shows signs of being uncomfortable when it moves. Your rabbit will put its foot down and then quickly reposition it. It will almost be like you trying to walk barefooted over sharp stones.

Treatment for Sore Hocks

- Remove the animal from its cage and examine its hock area. Are there areas where the fur has been worn off? Are these areas bleeding or infected? Examine the front feet, too. Sores can also develop on the front feet.

Turn your rabbit onto its back so that you can examine its hocks.

- Clean the hock area and apply a medicated cream, such as Bag Balm or Preparation H.

You can help your rabbit to recover more quickly by checking its cage for things that might have caused the condition, such as sharp spots on the cage floor or wire that has been put on upside down (see page 37). Place a board or piece of carpeting in the cage, so your rabbit can rest off the wire. Be sure the cage is clean, so infection will be less likely.

Sore hocks take a long time to heal, and if untreated, the spot can become infected. The sooner you start treatment the better your success rate will be.

Sunstroke/Heatstroke

You know how hot it can feel when the temperature rises into the nineties. Just think how this must feel if you are wearing a fur coat!

Hot weather can be very hard on rabbits. Extremely high temperatures can kill them. During a hot spell, it is wise to check your animals during the day. If the weather is hot, it is normal for a rabbit to stretch out in the coolest part of its cage. If your rabbit is breathing heavily and its muzzle is dripping wet, it may be suffering from heatstroke or sunstroke. You should act immediately in order to save the rabbit.

Treatment for Sunstroke or Heatstroke

- Move the rabbit to a cool location. The cellar in your house is probably several degrees cooler than outside temperatures.
- Wipe the inside and outside of the rabbit's ears with ice cubes. Because the blood vessels in a rabbit's ears are closer to the surface than those in other parts of its body, ice applied here will cool it faster than applying ice to any other spot. Wrap the ice in cloth, so that you can hold it more easily.

Causes of Sore Hocks

Some rabbits seem more likely than others to get sore hocks:

- Nervous rabbits that stamp their feet a lot may develop sore hocks more often than their calmer relatives.
- The soft, short fur of the Rex breed makes sore hocks more likely.
- Thinly furred foot pads may result in sore hocks.
- Some of the very large breeds seem to get sore hocks more often. These animals carry more weight, and therefore put more stress on their hock area.
- Animals whose feet are small compared to their body size are likely to get sore hocks.

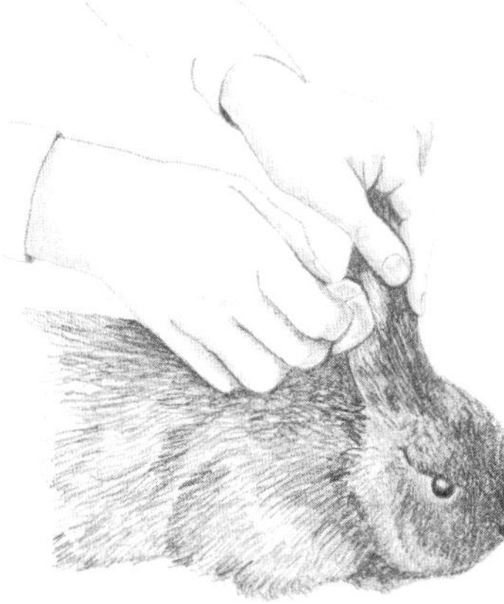

Wipe the inside and outside of your rabbit's ears with an ice cube to counteract heatstroke.

- *Special Care:* In advanced cases of sunstroke, the rabbit may be almost lifeless. In this situation, every second is important. To cool the rabbit as quickly as possible, gently place it in a container of water. Use water that is at room temperature, not cold water, so that you don't give it a sudden shock. Do not let the rabbit's head go under the water, just wet it up to its neck. Once it is wet, place it in a quiet, shaded area to recover.

Wool Block

Rabbits are very clean animals. During their normal grooming process, they naturally swallow some loose fur. Sometimes they swallow so much that a ball of fur forms in the digestive tract, and nothing in the rabbit's digestive system dissolves this fur. Because Angora rabbits have extra long coats, we see this condition more often in Angora breeds. Although we call this condition *wool block,* it can also happen to rabbits with other types of fur, including normal, Rex, and Satin.

If your rabbit is eating very little and is losing weight, wool block may be the problem. Check the rabbit in question. Are its teeth okay? Are there signs of diarrhea? Does the rabbit have a runny nose? If the answers to the above questions are no, then wool block is a good possibility.

An animal with wool block usually has a big lump in its stomach, so it thinks it is full and will not eat. This is why the rabbit loses weight, even though food is available. An animal with wool block will continue to drink, so treatment is usually done through its drinking water. The fruits pineapple and papaya contain a substance that can break down the fur balls. Feeding chunks of fresh pineapple is one common treatment for wool block. Adding fresh pineapple juice to the rabbit's water is a second. Health food stores sell papaya tablets that can be given to the rabbit with its feed.

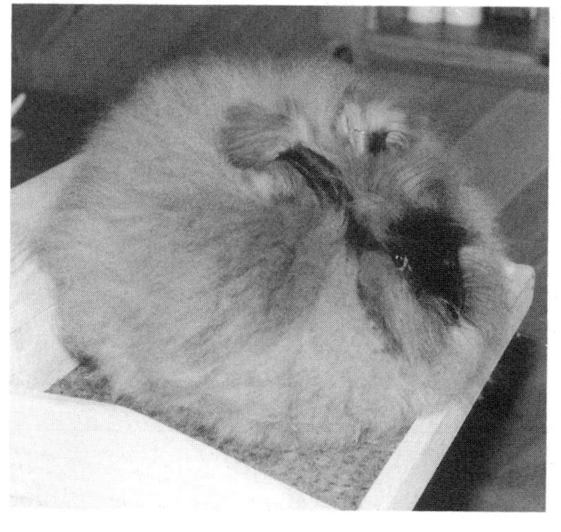

PAMELA T. BERNARDINI

Because of their long fur, Angora rabbits may occasionally suffer from wool block.

Treatment for Wool Block

- To insure that the pineapple or papaya mixture gets into your rabbit's system, use an eye dropper to hand-deliver a dose.
- As a preventative, many Angora breeders offer pineapple or papaya tablets on a regular basis.

Worms

Rabbits raised in wire cages very seldom have a problem with worms. If you see tiny worms in your rabbit's droppings, treatment is very simple.

Treatment for Worms

Piperazine, a product commonly used as a cat wormer, is used very successfully for rabbits. Normal treatment is to put one dose of liquid Piperazine in your rabbit's drinking water (1 teaspoon of Piperazine to 1 gallon of water). Treat for 3 to 5 days, wait 10 to 12 days, and then treat again for 3 to 5 days. Order Piperazine from a rabbitry supplier, or look for a cat wormer that lists Piperazine on the label.

Be Prepared

If you start with healthy rabbits and take good care of them, you should have very few health problems. If a problem does arise, early treatment will increase your rabbit's chances for a full recovery. Having the necessary things on hand will help you start treatment as soon as a problem is noticed. Why not assemble a Rabbit First-Aid Kit? You may already have some of the supplies for your first-aid kit in your home. Be sure to get a parent's permission before taking anything.

For those things you have to purchase, find several other rabbit owners who would also like to make a first-aid kit for their rabbits. Most medicines come in fairly large amounts. It helps if you can share the supplies and the cost with several others.

A Rabbit's First-Aid Kit

- ❑ *Several small bottles or jars.* Empty containers from prescription drugs work well.

- ❑ *Large container* to hold your first-aid materials. A school lunch box that is no longer used or a plastic shoe box is about the right size.
- ❑ *Labels* so you can mark each container with the following information:
  - Name of the medicine
  - What the medicine is used for
  - How the medicine is given
  - How much medicine is given
  - How often the medicine is given
- ❑ *Cotton balls* to clean wounds or apply medicines. Store these in a plastic bag.
- ❑ *Cotton swabs* to clean wounds or remove ear mite crust. Store swabs in a plastic bag.
- ❑ *Hydrogen peroxide* to clean wounds.
- ❑ *Disposable rubber gloves.*
- ❑ *Nail clippers* for keeping nails trimmed. These may also be used to trim teeth. Use ordinary nail clippers intended for humans, or if you have a large breed, use dog nail clippers.

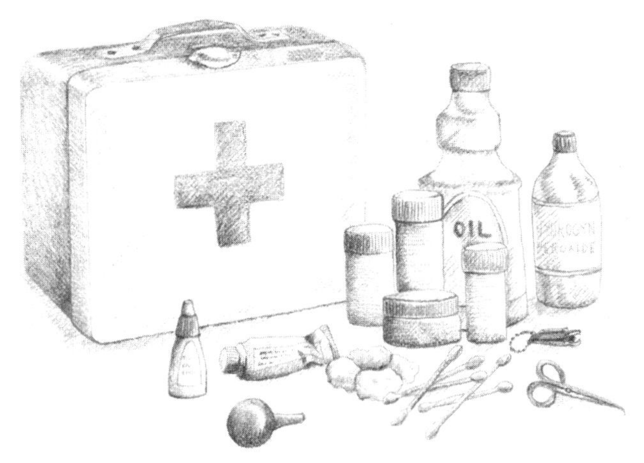

Be prepared for emergencies by keeping a fully stocked first-aid kit at hand.

- ❑ *Small pair of sharp scissors* to trim hair from around wounds.
- ❑ *Plastic eye dropper* to give liquid medicines or put oil in ear to treat ear mites. Wash well between uses. Store in a small plastic bag.
- ❑ *Miticide or cooking oil* to treat ear mites.

(The following products can be purchased at local farm or drug stores.)

- ❑ *Antibiotic cream* for wounds, and *Bag Balm or Preparation H* for sore hocks.
- ❑ *Papaya tablets* for prevention and treatment of wool block.
- ❑ *Piperazine* for worming.
- ❑ *Sulfaquinoxaline* for coccidiosis.
- ❑ *Eye drops* for nest box eye infections (page 89).

You and Your Veterinarian

You will become very capable to handle some health-related conditions, but some conditions require the skills of a veterinarian. Another valuable item in your first-aid supplies should be the name and telephone number of a local veterinarian. Since not all veterinarians have had a lot of experience treating rabbits, it will be worth the effort to find a person with the knowledge to provide the best veterinary care. Talk with other local rabbit owners. They may be able to direct you to a person who has been helpful with their rabbits. As rabbits become more and more popular as pets, veterinarians have more opportunities to treat them. This trend has made it easier for rabbit owners to find good veterinary care for their animals.

Breeding, Birth, and Care of Newborns

Producing your own stock is one of the most exciting parts of your rabbit project. Before you make plans to breed your rabbits, be sure to discuss this with your parents. You will also want to be sure that you have some potential markets for the young that are born. (Read chapter 9, "Marketing Your Rabbits," for some ideas.)

Breeding. *Raising rabbits.*

Once you decide to breed your rabbits, you should

- Decide when to begin your breeding program
- Make sure your rabbits are in good health
- Decide which rabbits to breed

When to Breed

Animals that are to be used as breeders must be mature and healthy. The larger the breed, the more time the rabbits need for their bodies to grow and become ready to produce young.

- Small breeds of rabbits should be at least 4 months old before their first breeding.
- The medium breeds—those that mature at 9 to 11 pounds—are generally ready to breed at 5 to 6 months of age.
- The giant breeds mature even later, and are not ready for breeding until they are about 9 to 12 months of age.

Signs That Your Doe Is Ready for Breeding

- The doe rubs her chin on her food dishes or on the edge of her cage. This rubbing is the way rabbits leave their scent to mark territory and to let other rabbits know that they are around.
- The doe's genital opening looks reddish. A pale or light pink genital opening indicates that the doe is probably not ready to mate.

Health Check

Producing young is work for the rabbit's body. That's why it's important for you to take time to check each animal's health before you consider breeding it. Breeders should be in good physical condition, not too fat and not too thin, and free of signs of disease, especially coccidiosis.

Selecting Breeding Pairs

If your rabbits are the proper age and are in good health, your next step will be to decide which rabbits to breed. If you are starting with one buck and one doe, then the decision is made for you. If you have several different bucks and does, you will want to choose the two rabbits that have the greatest potential for producing good-quality babies. This is a time when knowing as much as possible about your breed can really pay off. A good-quality buck and a good-quality doe are more likely to pass on good characteristics than poorer-quality breeders. If one of your rabbits has an area that you would like to improve upon, try to select a *mate* that is especially strong in that area. For instance, if your doe would have better type (the shape and proportions of an animal) if her shoulders were broader, you should look for a buck that has good broad shoulders. Such a mating is more likely to produce young with

Mate. *Partner in reproduction.*

better shoulders.

Once these decisions are made, you are ready to begin your breeding program.

Putting the Doe and Buck Together for Mating

Once you have selected the two rabbits that you wish to breed, the next step is to have the breeding take place. Bucks are almost always willing to breed. Does sometimes resist.

It is very important that your rabbits are not left unattended during mating. You need to be there to know if mating takes place. You also want to be there to separate the animals if problems arise.

When you are ready to put the male and female together for breeding, always bring the doe to the buck's cage. Rabbits are *territorial.* If the buck is put in the doe's cage, the doe may be more involved in defending her territory from the buck than in mating with him. By taking the doe to the buck's cage, you avoid arousing her territorial *instinct,* and she is more likely to mate.

Instinct. *Natural tendency to behave a certain way.*

A buck and doe preparing to mate.

When a doe and buck are placed together, the buck is usually ready to mate with the doe and chases after her. If the doe is ready to mate with the buck, she will not run away and will raise her hindquarters after the buck starts the mating motions. To hold his position on the doe, the buck grabs a mouthful of her fur. Sometimes the buck pulls out some of the doe's fur, but he does not usually hurt her. You will be able to tell when mating has occurred because the buck will make a noise (usually a groan or squeal) and then he will fall off to the side of the doe.

Sometimes the doe does not seem to want to mate with the buck. She will not raise her hindquarters, and she will run around the cage to try to get away from him. She may change her mind and be ready to mate after a few minutes, so leave them together for a short while. However, if the doe does not show interest within ten minutes, take her back to her own cage and try again in another day or two.

Occasionally, a doe is not ready to mate and will fight with the buck. If the buck and doe fight, separate them and try again another day. You may want to have an adult nearby to help you remove the doe when this happens. If the doe is upset, she may bite! Be prepared: Have a pair of heavy gloves nearby in case you need to remove the doe quickly.

Sample Hutch Card

Doe's name ______________

Doe's number ______________

Buck's name ______________

Buck's number ______________

Date litter is due ______________

Number born ______________

Number weaned ______________

Comments ______________

____

____

____

After Mating Has Occurred

Once the mating has taken place, your first job is to return the doe to her own hutch. The next important thing to do is to write down when the doe will *kindle* (give birth). You can expect the litter to be born in 28 to 35 days. The average time is 31 days. On a calendar, count 31 days ahead and you will know when to expect the litter. Having a calendar in your rabbitry is a handy way to keep track of upcoming litters. It is also a good

indle. *When rabbits ive birth.*

idea to keep this information on the doe's cage using special cards called *hutch cards.* Hutch cards have space to record information about the doe and the babies she produces. Pre-printed hutch cards are often available from feed companies, or you can make your own (see page 80). A hutch card should list

- ❑ Doe's name or number
- ❑ Name or number of the buck used
- ❑ Date when the litter is due
- ❑ Number of young born
- ❑ Number of young weaned
- ❑ Comments

If your hutch is outside, slip the card into a plastic sandwich bag, so the weather can't erase this important information.

Care of the Pregnant Doe

Continue to provide good basic care to the bred doe. Proper amounts of food and water in clean housing are more important than ever.

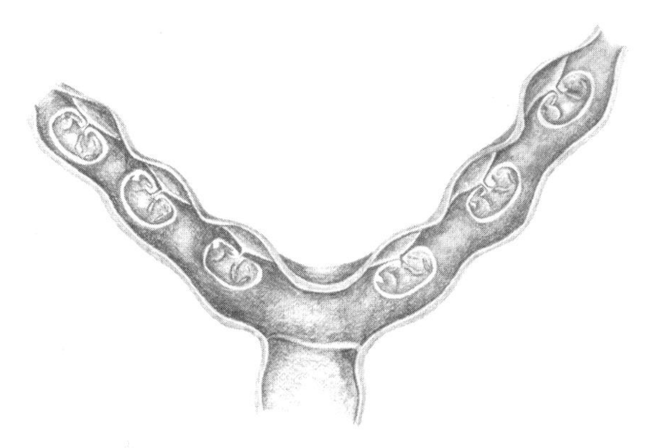

A cross-section of doe's uterus, showing developing rabbit fetuses.

Feeding Pregnant Does

Overweight does have more problems, so resist the urge to give extra feed to the doe during the early stages of pregnancy.

- Most breeders keep the doe on her normal ration for the first half of the pregnancy period (fifteen days).
- During the second half of the pregnancy, increase the doe's feed gradually. (See the chart on page 55 for guidelines.) Just before a doe kindles, she will eat less, and you can decrease her feed during the last 2 days.

Getting Ready for Kindling: The Nest Box

Your doe will need a special place, or nest box, in which to have her litter. Nest boxes come in a variety of sizes and types. The size of the box increases as the size of the doe increases. Nest boxes can be constructed from wood, wire, or metal.

Nest box guidelines

- Small breeds: 14" long, 8" wide, 7" high
- Medium breeds: 18" long, 10" wide, 8" high
- Large breeds: 20" long, 12" wide, 10" high

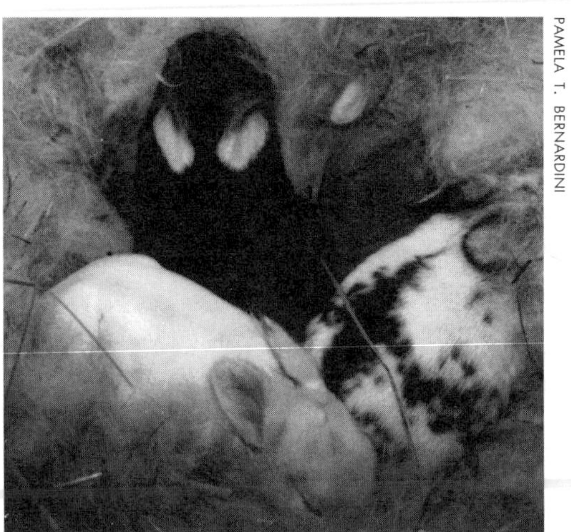

PAMELA T. BERNARDINI

A nest box is a protected area for newborn rabbits.

Wood Nest Box

Homemade nest boxes are commonly made from wood. An adult will probably have to cut the pieces out for you, but you should be able to assemble the nest box yourself. Use ⅜-inch plywood. Many does love to chew on the edges of the nest box, and plywood is a hard type of wood that holds up well. Use wood glue and nails to assemble the nest box. Drill a few holes in the bottom of the nest box. This allows for drainage, which helps keep the box from getting too damp. Damp nest boxes can contribute to diseases in young rabbits.

Some people like to have a cover over part of the nest box and some do not. A top helps to keep heat in, so it is recommended in colder climates. Another advantage is that some does seem to enjoy resting on the top of the nest box. A disadvantage of the partial cover is that it also keeps moisture in.

Although wooden nest boxes are widely used, they need to be thoroughly disinfected between litters. Use a chlorine-bleach-and-water solution (1 part household bleach to 5 parts water) to clean the box and let it dry in the sunlight. Store nest boxes where they won't be contaminated by other animals, such as dogs, cats, or rodents.

Wire Nest Box

You can also build nest boxes from ½" x 1" wire. These boxes are lined with cardboard in cooler weather to prevent drafts. You can make your own liners, or purchase them. Cardboard liners are thrown away after each litter. This makes it easier to keep wire nest boxes clean.

Metal Nest Box

The third type of nest box is made of metal. These boxes can be purchased in a variety of sizes. An advantage of metal nest boxes is that they are very easy to clean. During cold weather, many breeders use cardboard liners for insulation in this type of nest box. You can also set the box on Styrofoam or a thick layer of newspapers.

Nesting Materials

All types of nest boxes need nesting materials for the doe to use to build her nest. Materials vary from location to location, but hay or straw is the most common. I generally use about 2 inches of wood shavings in the bottom of the nest box and add lots of clean hay on top. Use less bedding in the summer. In winter, place layers of cardboard in the bottom of the

Putting the Nest Box in Place

Whatever your choice of nest box and bedding materials, be sure the prepared nest box is placed into the doe's cage on the 27th or 28th day after mating. If you put the box in too early, the doe may soil it with droppings or urine. Any later may be too late for a doe that kindles early. Be sure you do not place the nest box in the part of the cage the doe has chosen as her toilet area.

box followed by wood shavings, with hay packed in tightly on top.

Normal Pre-Kindling Behavior

As the big day approaches, you will notice some behavior changes in your rabbit. Sometime before she kindles, your doe will prepare a nest. Some does jump into the nest box as soon as it is placed in the cage. They will busily dig and rearrange the bedding to suit themselves. You may see the doe carrying a mouthful of hay around the cage as she decides where to place it. When you see the doe carrying mouthfuls of bedding, you can be sure that kindling time is near. She also uses fur that she pulls from her own chest and belly for nest material. This behavior not only provides soft nest material, but also exposes her nipples, making it easier for her babies to nurse after they are born.

One reason to look at the new litter is to see if there are any babies that did not survive. Dead babies and soiled bedding should be removed as soon as possible.

Some does, like some people I know, leave everything to the last minute. They show little interest in the nest box. You may feed your doe one evening and see an untouched nest box, but the following morning it is all over: the nest is made, fur is pulled, and babies have already arrived!

Another change you will notice is that the doe will eat less the day or two before kindling. This change in appetite is normal before kindling. A third change you may observe is that your doe may act more nervous than usual.

Checking on the New Litter

You will be eager to see if your doe has had a successful litter, and it's important to check on the well-being of both the young and the doe. Remember that it is also still important to keep the doe's area quiet. Some does do a very neat job of kindling, and you may not even notice that the litter has arrived. If you look closely, you will see that the fur in the nest appears all fluffed up. When you observe more carefully, you may notice that the fur seems to move.

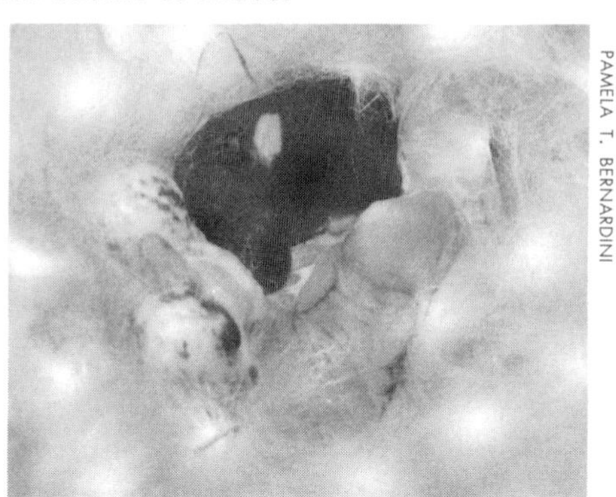

PAMELA T. BERNARDINI

Baby rabbits cuddle together for warmth.

Be sensitive to the doe at this time. If she appears nervous, you will want to distract her before you try to look at the litter. A nervous doe will often jump into the nest to protect her babies. Sometimes babies can be stepped on and injured. Bring a handful of fresh hay, a slice of apple, or another of the doe's favorite treats if she needs a distraction. Once she is occupied, you can check things out. Gently place your hand into the fur nest. You should feel warmth and movement. If the doe is still calm, part the fur and take a peek. You should see a group of several babies cuddled together for warmth. Newborn bunnies do not have fur and their eyes will be closed. Take a quick count. This may be a challenge as the bunnies will be overlapped and piled on top of each other.

Quiet Please

Try to keep things quiet around your doe at kindling time. This is not the time to invite your friends over to see your rabbits. The presence of dogs or wild animals such as rats or raccoons can also disturb the doe. Even changes in weather, such as a sudden thunderstorm, can upset the doe.

Fostering

A medium-sized doe usually has eight nipples from which her young will nurse (smaller breeds have fewer nipples). If a doe has more than eight babies, some of

Foster. *Have a mother rabbit accept as her own a baby rabbit that she did not give birth to.*

the babies are not going to get as much to eat as the others. To prepare for this sort of a situation, many rabbit breeders breed more than one doe to kindle at the same time. Then if a doe has an unusually large litter, some of the babies may be moved into the nest box of a doe that had a small litter. This is called *fostering.* Most does will accept the additional babies if they are about the same size as their own and if they smell like their own babies.

Some people go to great lengths to trick the doe into thinking that these additional babies are her own. One method is to put a strong-smelling substance on the nose of the doe. The idea is that a little dab of perfume or vanilla makes everything smell the same to the doe, and therefore she will not notice the extra babies.

Tip for Successful Fostering

I have always successfully fostered babies without using special scents. Here's how: Rabbits nurse their young only at night. If I need to move babies to another nest box, I make the move in the morning. The babies cuddle in with their new littermates, and by feeding time they all smell the same to the doe.

Care of the Newborns

Your doe has kindled her litter. In the majority of cases, all has happened normally, and you now have a healthy litter to raise. Of course, the doe will continue to do most of the work, but you can do several things to help those young grow into healthy rabbits.

The Doe's Needs

Keeping the new mother in good condition is the most important thing that you can do for the babies. Feed the doe a limited amount of feed for the first few days after kindling. She will continue to eat and drink, but you will notice that she eats a little less than her normal amount. After a few days her appetite will increase. She will then require an increased amount of feed to meet the needs of her own body and produce milk for her growing litter. The doe's need for water will also increase at this time. Always have clean fresh water available to a doe that is *nursing* a litter.

Nurse. *Suck milk from the mother's nipple.*

Checking the Litter

You will want to continue to check the litter every 2 or 3 days. It is normal for the young to react to a human hand invading their nest, so don't be surprised if the bunnies jerk away from your touch. Just be careful that they do not injure themselves. Do not let individuals squiggle away from the warmth of the group.

Occasionally, you may find that a baby has fallen out of the nest box. Mother rabbits do not carry their babies like cats carry their kittens. If you find a young bunny out of the nest box, be sure to place it back with its littermates. In addition to needing the warmth of the group, babies that stray from the nest are likely to miss mealtime. The mother will nurse only one group of babies. Young that are separated won't get fed and will soon die.

In cold weather, young that leave the nest too early are at particularly great risk of exposure. Remember, rabbits are born without fur, so if they leave the warmth of the nest, they soon become chilled. On cold days, it is always wise to begin your morning chores with a quick check for straying babies. If you find a baby outside the nest and it still feels warm to the touch, it can simply be placed back with its littermates. If the baby feels cold to the touch, it may need some extra help regaining its body temperature.

Baby rabbits begin to grow fur within a few days of birth, and by 2 weeks, they will be completely furred.

Hand-Feeding a Rabbit

Sometimes a doe dies after her young are born. When this happens, you may wish to try to feed and care for the babies until they are mature enough to take care of themselves. To feed the young, make up the following mixture and store it in the refrigerator:

Methods of Warming a Chilled Bunny

- Take the chilled youngster into a warm place.
- If you have a heating pad, turn it on, cover it with a towel, place the bunny on it, and place the pad and bunny in a small box. Make sure the box provides protection if you are bringing the bunny into an area where there are pets, such as dogs or cats.
- Share your body heat. If you tuck the baby inside your shirt, it can rest against you and gain warmth. This will feel like putting an ice cube down your shirt! But it is a wonderful experience to feel a cold, almost lifeless bunny slowly regain consciousness.

- 1 pint skim milk
- 2 egg yolks
- 2 tablespoons Karo syrup
- 1 tablespoon bonemeal (available in garden supply centers)

Genitals. Reproductive organs, especially those on the outside of the body.

Use an eyedropper or straw to feed this mixture to the babies twice a day. Feed them until they stop taking the milk (usually about 5 to 7 cc). In addition to keeping the young warm and fed, you must also be sure that they urinate and defecate regularly. To do this, simply stroke their genitals with a cotton ball after you feed them. Follow these procedures until the young are 14 days old.

Keep the Nest Box Clean

Most does will enter the nest box only at feeding time. Once in a while, however, you may have a doe who spends more time than necessary in the box, especially when the nest box is too big. This often leads to a dirty and damp nest box—a perfect setup for disease! If you see droppings in the nest box or if the bedding materials feel damp, you should clean the nest box before problems begin. Remove the nest box from the doe's cage. Remove and save any clean and dry fur that the doe has pulled. Place this fur in a container—a cardboard box or a clean bucket with some clean bedding in the bottom—and gently place the litter into this temporary home. Now, clean out the soiled bedding and replace it with clean bedding. Use your hand to form a hole in the clean bedding. Line this area with some of the saved fur, place the bunnies into it, and cover them with the remaining fur. Return the clean nest box and litter to its location in the doe's hutch. Your bunnies now have a clean and healthy nest to grow up in.

Watch for Eye Problems

Your bunnies should open their eyes at 10 days of age. It is important to watch for this. Sometimes a bunny may not be able to open its eyes even though it is old enough to do so. This usually indicates an eye infection. An infection may affect one or both eyes. Eye infections may happen to one or several littermates. Tending to this condition as soon as possible is your best means of avoiding blindness.

A bunny that is unable to open its eyes usually has a dry, crusty material sealing the lids together. Use a soft cloth, cotton ball, or facial tissue soaked in warm water to wash and soften the crusted eyelids. Once softened, you can use your fingers to gently separate the eyelids. Gently wash away any remaining crusty material. If the edges of the eyelid look reddish, I place a drop of over-the-counter eyedrops for people (such as Visine or Murine) in the eye.

You will need to check the bunny for several days and repeat this procedure as necessary. Once the eye is cleaned and opened, most bunnies recover very quickly.

Sometimes when you open an eye as directed above, there will be a white discharge. This pus indicates a more serious condition. After gently washing away the crusty material and pus, special eyedrops (containing the antibiotic neomycin), available from your veterinarian, should be used. You will need to clean the eye and apply medication for several days, when the infection should clear up.

Some rabbits have a normal-looking eye after recovering from an infection of this type. However, some may develop a cloudy-looking area on part or all of the eye. This usually means that the rabbit will be partially or totally blind in that eye.

Beginning to Handle the Bunnies

As the bunnies approach 3 weeks of age, they will begin to come out of the nest box. At this age, they are also able to find their way back into the box, so you don't have to worry about constantly watching for strays. The bunnies will begin to eat pellets and drink water. This will mean you need to have lots more food and water available. Of course, the bunnies are still getting milk from their mother. Bunnies will continue to nurse until they are about 8 weeks old.

This time period, from 3 to 8 weeks, is when the young rabbits will be at their cutest. You will enjoy watching them play together and explore their world. This is an excellent time to begin to handle the young rabbits. Be especially careful when you pick them up for the first time. Young rabbits are easily frightened and may struggle to get away. Remember to wear a long-sleeved shirt to avoid scratches. Although young rabbits may be squirmy at first, they adapt very quickly. Try to spend a few minutes each day handling the bunnies. Rabbits that are handled properly as youngsters will be more easily handled throughout their lives.

Sexing the Litter

Later, when you wean the litter (pages 92–93), you'll want to separate the bunnies by sex. In order to do that, you have to be able to tell the difference between the two sexes. Making this distinction is called *sexing.* Although mature rabbits are easy to sex, it takes some practice to sex young rabbits accurately. Try to sex the litter around 3 weeks of age, when you are beginning to handle the youngsters on a regular basis.

Part of the bunnies' handling experience should include getting used to being turned over for examination, which is necessary for sexing.

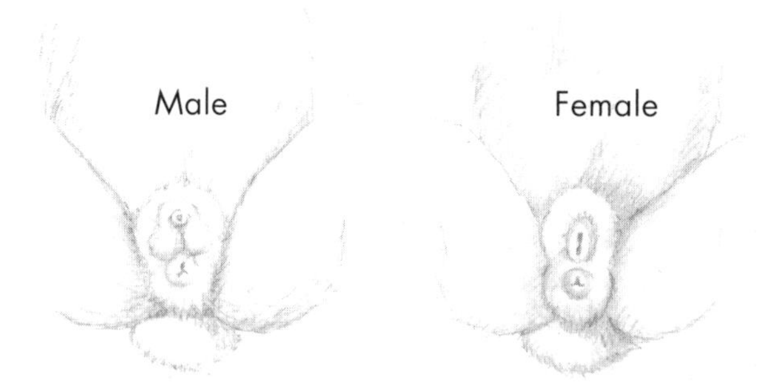

Compare the genital area of your baby rabbits to this drawing of male and female bunnies.

The sexual openings are quite small on young rabbits, and it can be difficult to be absolutely sure which sex an animal is. To help you gain confidence in sexing your litters, start to check them at an early age and re-examine them weekly. Use a black marking pen to write a "B" in the ear of those you believe are bucks and a "D" in the ear of those you believe are does. Each time you re-examine the litter, check to see if you agree with what you thought when the rabbit was a week younger. You may find that you will change the letter in the ears of several animals. If you continue to check through weaning, you will find that sexing young rabbits gets easier as the rabbits get older. Even experienced rabbit breeders can make a mistake sexing young rabbits, so don't get discouraged.

Keeping the Hutch Clean

As the litter grows, it will become more of a challenge for you to keep their hutch clean. If you are using crocks to feed and water the litter, you will notice that these containers often become soiled, as young bunnies climb into the crocks to reach the feed and leave their droppings behind. Feeding smaller amounts more often will save more feed than feeding a large amount once a day. Several small feedings will also give you more opportunities to clean out the feed dishes. The

How to Sex Young Rabbits

- One of your hands restrains the rabbit's head.
- Place your first and second fingers of the other hand around the base of the tail. Use your thumb to press down gently in front of the sexual organ.
- If the rabbit is a doe, you will see a slit-like opening. This opening will begin near your thumb and then slope down toward the rabbit's tail.
- If the rabbit is a buck, the opening will look rounded and will protrude slightly.

cleaner their environment, the healthier your litter will be.

Removing the Nest Box

Now that the bunnies are out and about, you may wonder if they still need their nest box. Because the bunnies are eating pellets, they are also producing droppings that make the nest box dirty. Unless the weather is very cold, it is best to remove the nest box when the litter is 3 weeks old. If you choose to leave the nest box in the cage longer, be sure to clean it and replace the bedding whenever it becomes soiled.

Weaning and Separating the Litter

Weaning. *Changing the way a baby is nourished from nursing to eating other feed.*

As the litter reaches the age of 5 to 8 weeks, the young rabbits are ready to be *weaned.* Weaning is the process of removing the young from the mother so they are no longer receiving nourishment from her milk. By 8 weeks, the young rabbits are used to eating pellets and drinking water. And after 8 weeks of nursing the litter, the doe's body needs a rest. Producing milk is work for the rabbit's body, and the doe needs a break before she is ready to raise another litter.

Also, by 8 weeks of age the littermates begin to behave like adult rabbits in some ways. Around this time, a group of youngsters that has been living happily together begins to have mature instincts about having their own territory and breeding. Young bucks will begin to chase after their sisters. Rabbits can mate and produce litters before they are full grown. Having a litter at a young age is very stressful for a doe, so part of the weaning process includes separating the litter by sex.

Many breeders feel that the weaning process is easier on the mother if not all the young rabbits are removed at once. Since most litters are about half bucks and half does, I suggest that the young bucks be separated first. A few days later, the young does can be moved. This two-part weaning process allows time for the doe's body to adapt gradually to the need for reduced milk production.

Weaning time is a good time to attend to three other tasks that need to be accomplished with each litter: 1) culling, 2) tattooing, and 3) writing pedigrees.

Culling. Culling is a selection process. This will be your opportunity to decide which animals in the litter are best suited for show animals, breeders, pets, or for other purposes. When you cull the litter, you will be using the same skills and knowledge that are used in judging or in selecting an animal to purchase. You will want to

- Check the health of each animal
- Check to see if there are any eliminations or disqualifications present (see page 15)
- Evaluate each animal as to how closely it meets the standards set for its breed

The youngsters that do well in all three areas become your show prospects, future breeding stock, or stock to sell as purebreds. Rabbits that are healthy, but not quite so outstanding as examples of their breed, most generally become pets, or, if of the medium or larger breeds, they may be sold as meat rabbits. If you discover any health problems, these animals should be caged away from the others, and they should receive proper treatment. (See chapter 6, in which health issues are discussed.)

Once you have selected the best animals from the litter, it is best to house these rabbits in individual cages if possible. As rabbits mature, they do best if they have their own private space. Rabbits housed together often fight to decide who will be the boss of the hutch. Although rabbit fighting seldom leads to life-threatening injuries, torn ears, scratches, and bite injuries are

Removing the Nest Box

I often allow the nest box to remain a little longer, more for the doe's sake than for the litter's. As the bunnies explore their hutch, they keep after Mom for extra attention. Since my nest box has a partial top, it provides a place for the doe to get away from the demands of the litter. Once the babies are big enough to climb on top of the box, the doe's hideaway no longer serves its purpose, and I remove the nest box.

Culling. *Separating rabbits according to intended use — for example, show, breeding, pet, meat.*

common results. Young rabbits being raised for pets or meat can be kept more than one to a cage, but the cage should provide adequate space, and each cage should house just bucks or just does. Rabbits can be kept more than one to a cage until they are sold or up to 4 months of age. Be alert if any fighting occurs and separate those involved.

Tattooing. Tattooing is generally done at weaning. Once a litter is split up into several cages throughout the rabbitry, it can become difficult to remember who is who. Rabbits that you plan to keep to sell as purebreds should be tattooed. You do not have to tattoo pet stock or those destined for the meat pen. See pages 123-27 for information about how to tattoo a rabbit.

It is also a good idea to attach a card to each cage to identify the rabbit housed there. This card should include information such as when the animal was born, what sex it is, and who the parents are. Having this information handy will make it easier to show your rabbits to prospective buyers.

Pedigrees. You should write pedigrees for all stock that will be kept for breeding and that you hope to sell as purebreds. As with tattooing, you need not write a pedigree for pets and meat animals. Writing pedigrees takes time, and it is easy to put this task off. Try to complete your pedigrees as part of the overall weaning process. See pages 127-30 for information about how to fill out a pedigree.

Showing Your Rabbit

Why Show Rabbits?

If you enjoy raising rabbits, chances are you will enjoy showing rabbits. Showing is not a requirement of being a rabbit raiser, but it can be one of the most interesting and fun parts. Let's look at some of the advantages and disadvantages.

Advantages:

- Rabbit shows are fun. You get to go places and meet new people. You may even win some prizes!
- Shows give you opportunities to learn more about your rabbits and about rabbits in general. Trained, experienced judges give their opinions of your rabbits. You get to see how your animals compare to others of the same breed.
- Showing puts you and your rabbits in the spotlight. Shows are a good place to advertise and sell rabbits.

Disadvantages:

- Taking your rabbits to shows can be stressful for them, and you may be exposing them to other rabbits that are sick.

PAMELA T. BERNARDINI

Rabbit shows are a good place to learn about rabbits in general.

- Showing takes time. Most shows mean early-morning travel, a full day at the show, and then the trip home. You need a cooperative adult who is willing to provide transportation and to spend the day at the show.
- Showing costs money. Most shows charge an entry fee for each rabbit. Travel and food can add to the total show expenses.

Most of the breeders I know feel that the advantages of shows outweigh the disadvantages. There are also ways you, as a breeder, can eliminate or reduce the negative aspects of showing. Some simple ways to do this are:

- Enter only those rabbits that are in the best of health.
- Don't show very young animals or does that have recently kindled.
- Keep show rabbits in a separate part of your rabbitry. Carefully watch for health changes after a show.

- Provide comfortable travel conditions for each rabbit.
- Before entering a show, check each rabbit for eliminations or disqualifications. Don't waste money entering animals that the judge will not place.
- Carpool with friends and pack a lunch to help keep travel expenses lower.

Entering a Show

Rabbit shows come in all sizes, from small gatherings sponsored by local 4-H Clubs to the annual American Rabbit Breeders Association (A.R.B.A.) National Convention and Show.

Show Book

Each show has a secretary or person in charge of receiving and organizing the show entries. To enter a show, you must first write to the show secretary and ask to have a show book or show catalog mailed to you. This book will answer important questions about the show, such as:

- When and where will the show take place?
- Is the show *sanctioned* by (following the rules of) the A.R.B.A.?
- How much is the entry fee?
- When are entries due?
- Are the entered rabbits kept overnight?
- Are they kept in show carriers or are cages provided?
- What other rules apply?

Sources of Information about Rabbit Shows in Your Area

- The American Rabbit Breeders Association
- A local A.R.B.A. club
- Your county 4-H office
- Local rabbit breeders who show their rabbits

Entry Form and Fee

The show book also includes an entry form for you to fill out. Most shows require that you send your entry form to the show secretary by a date that is listed in the show book. Some shows have you bring the entry form with you on the day of the show. Whatever the system, a neatly written entry is a must. Be sure to list all the necessary information for each rabbit that you are entering.

RABBIT SHOW ENTRY & REPORT

Entry No. ______ Date 8-10-92

Exhibitor Name Chris Jackson Rabbitry Name chris's critters

Address Main Street City Springfield State MA Zip 01234

ARBA No. Open ______ Youth JACKCHOU Specialty Club No./s ______

State Assn. Open ______ Youth ______ Ribbons ☑ Yes ☐ No

					For Secretary's Use Only				
Coop	Breed & Variety	Ear Number	Sex & Class	Entry Fee	No. In Class	Place	Points	Cash	Specials
112	Dutch-Black	J 17	Sr buck	1 50	4	2	16		
113	Dutch-Blue	J 14	Sr buck	1 50	5	3	15		
114	Dutch-Blue	J 12	jr buck	1 50	4	1	24	3	B.O.V.
115	Mini Lop BIC	J 9	sr buck	1.50	12	3	36		
116	Mini Lop BIC	J 6	jr buck	1.50	10	4	20		
117	Mini Lop BIC	J 4	jr buck	1 50	10	2	40		
118	Californian	J 32	sr due	1.50	6	2	24		
119	Californian	J 21	int buck	1.50	3	2	12		
120	Californian	J 57	jr buck	1.50	5	1	30	5	B.O.B.
121	Californian	J 2	jr buck	1.50	5	3	15		

Total Entry 15.00 Total Cash $8.00

Display Awards

1. ______
2. ______
3. ______

Total Points 232

Signed Sue Secretary
Secretary

Agway Inc., Country Foods Division
PO Box 4933, Syracuse, NY 13221
AGWAY

The show entry form that you fill out will be returned to you after the show with the show results recorded.

- *Breed.* Your rabbit will compete against others of the same breed, so the show secretary must know what breed your animal belongs to. A.R.B.A.-sanctioned shows are open only to purebred rabbits. Fair shows and 4-H shows sometimes include competition for crossbred rabbits.
- *Group or variety.* Many breeds are subdivided into color groups or varieties. The book *Standard of Perfection* describes how your breed is subdivided. If your rabbit's breed has more than one group or variety, as most breeds do, you must list it on the entry.
- *Sex.* Is your rabbit a buck or a doe? Your rabbit will be placed into a class with animals of the same sex.
- *Class.* Show competition is further subdivided based upon the age of the animals. If your breed matures to an ideal weight of less than 9 pounds, your rabbit is shown either as a Junior (under 6 months of age) or as a Senior (6 months of age and over). If your breed matures to an ideal weight of 9 pounds or over, it can be shown as a Junior (up to 6 months of age), as an Intermediate (6 to 8 months of age), or as a Senior (8 months of age and older). Also, certain breeds have weight limitations on each age classification.
- *Tattoo number.* All A.R.B.A. shows require that a personal identification number be tattooed in your rabbit's left ear. (The right ear is reserved for the registration tatoo.)

Most shows require that you mail your entry fee along with your entry form. Since checks are safer and easier to mail than cash, you will want to ask an adult to help with this.

Be sure to make a copy of your entry form before you mail it. When the show date arrives, this copy will help you remember which animals you entered in the show.

Enhancing Your Rabbit's Condition

- Clean rabbits come from clean cages. Make sure there is no manure build-up in your rabbit's cage. Time spent cleaning will pay off on show day, especially if your rabbit is white.
- Remove loose hairs from your rabbit. You can use a soft brush, or simply dampen your hand and stroke the coat to remove loose hairs. If your rabbit has a lot of loose fur, it is probably in a *molt*. Molting is the natural process of growing a new coat. Grooming will help a molting rabbit look better, but a molting rabbit will be at a disadvantage at a show.

Getting Ready for a Show

After you mail your show entry, there is lots to do before the show day arrives.

Fitness for Showing

Check each rabbit for possible eliminations or disqualifications. Although the rabbit may have previously passed your checkup with flying colors, things do happen and conditions can change. Some common problems that might arise are a chipped or broken tooth, a missing toenail, or a change in weight. If you find a problem, you can help the rabbit recover by offering special care. If you have other rabbits, finding a problem before show day also allows you to choose a replacement animal to take to the show. Most shows allow you to substitute another of your rabbits that is the same breed and class.

Handling

Spend time handling your rabbit. The judge will find it easier to evaluate your animal if it is used to being handled and posed. The time you spend helping your rabbit get used to being handled will make show day easier on you, your rabbit, and the judge.

Conditioning and Grooming

Have your rabbit looking its best. Although the overall condition of the animal is not as important to the judging as body type, color, or markings, it can make a difference, especially in a close competition. Condition reflects the long-term care that the rabbit receives, and it can't be changed in a short time.

Tattooing

Be sure your rabbit has a legible tattoo in its left ear. Imagine how confusing it would be to have twenty New Zealand White Junior does on the show table at the same time! Since it is sometimes very difficult to tell one rabbit from another, a permanent tattoo insures that each animal is returned to its proper owner. (See pages 123–27 for information about tattooing.)

> Avoid using a cardboard box to transport your rabbit. Even with air holes cut on the sides, a box can overheat. On a warm day, a box can quickly become a dangerous place for your rabbit to be.

Travel Arrangements

Prepare comfortable travel arrangements for your rabbit. Many first-time exhibitors use whatever is available as a carrier: old bird cages, pet carriers, even cardboard boxes. Although these may appear to work, they are far from ideal. Pet carriers developed for other animals may be the right size, but they do not keep the rabbit away from its droppings, and the position of the door makes it difficult to remove the rabbit.

Special carriers designed especially for rabbits are available for purchase in a variety of sizes and styles, or you can build your own. Rabbit carriers are constructed of wire mesh and are designed so that droppings fall through the wire floor into an attached pan below. This type of carrier provides lots of fresh air for the rabbit, and keeps the rabbit safe and clean. Some carriers are subdivided to hold several rabbits. All-wire carriers are the best way to transport your rabbits. (See pages 44–46 for information on building carriers.)

Prior to each show, remove and clean the bottom tray of the carrier. Place wood shavings or other absorbent material in the clean tray. I also like to place some fresh hay in the carrier itself. This gives the rabbit something to snack on, and I don't have to worry about packing pellets and food dishes. Rabbit shows are usually daytime events, and since rabbits do most of their eating at night, you don't have to worry about upsetting your rabbit's feeding schedule.

If you are traveling to a show during hot weather, a small water bottle can be attached to the carrier. Another way to provide some nonspill moisture to traveling rabbits is to place a slice of fresh apple or carrot in the compartment. Many exhibitors also take a crock to provide water at the show. Taking along some water from home in a recycled milk jug is a good idea. Water from home will have a familiar taste, and your rabbit will drink it more readily. When you take more than one rabbit to a show, during the day you can place a crock of water with one rabbit for a period of time and then move it on to refresh another.

Where you place the carrier during the trip to a show is as important as the carrier itself. The open back of a pickup truck may be fine on a mild day, but rain or direct sun can be harmful to your rabbit. The back of a covered, well-ventilated pickup is a better way to go. Rabbits will also travel very nicely right in the cab of a truck or in a car with you. If you have a carrier with a solid tray that collects droppings, there will be little or no mess. When you travel together, you will be right there if any problems arise during the trip.

Do not Transport Your Rabbit in a Car's Trunk

- The closed trunk can become extremely hot.
- Dangerous engine exhaust fumes can seep into the trunk.
- Insufficient fresh air in a closed trunk can be harmful to your rabbit.

When You Arrive at the Show

Even with the best preparations, show day can still be hectic and confusing. Your day will go more smoothly if you arrive on time. This means allowing enough travel time so you arrive during the check-in period listed in the show book.

Upon arrival, your first challenge will be to seek out the show secretary or the person responsible for checking in exhibitors. You'll usually find this person seated at a table, surrounded by papers and a crowd of exhibitors. Take your place in line and await your turn. The proper line for you will vary from show to show. Some shows have one check-in for everybody. Some larger shows subdivide check-in, with one line for open (adult) entries and another for youth exhibitors. If this is a day-of-entry show, check-in is where you hand in your entry form and pay your entry fee. If you entered the show by mail ahead of time, check-in is when you can make last-minute changes to your entry paperwork. Be sure to let the show official know if you have any *scratches*—animals that you entered but did not bring. The show official needs this information so these animals can be removed from the show paperwork. If you are substituting a different animal for one you had previously entered, be sure to provide the correct tattoo number of your substitute. Recording changes like these helps keep the show running smoothly.

At check-in you will receive a copy of your entry or a *coop card* for each rabbit you have entered. Coop cards list information for each rabbit and assign it a *coop number.* This very important number becomes the animal's identification number for the show. At many shows, before your rabbit can be judged, you must write the assigned coop number in its right ear. Use a permanent black marker for this.

RABBIT SHOW

COOP NUMBER: ____________

ENTRY NO. ________ EAR NO. ________

BREED ____________________

CLASS: JR. ____ 6-8 MO. ______ SR. ______

BUCK _____ DOE _____ DOE and LITTER _____

MEAT ________ FUR CLASS ________

OWNER ____________________

ADDRESS ____________________

CITY ____________ STATE ________

SALE PRICE $ ____________________

A coop card lists information about your rabbit and assigns it a coop number.

Judging: Where and When?

Your next challenge is to figure out where and when your breed will be judged. Most rabbit shows make it the exhibitors' responsibility to bring their animals to the show table at the proper time, so it's important for you to be prepared. Finding out where and when your breed will be judged is easy at a small show that has only one judging table. However, many shows have several judges to evaluate all the entries. In that case, you'll need to walk around the show room and check the signs at the various judging tables. At a well-organized show, the sign at each judging area will list the breeds in the order in which they will be shown. Use these listings to plan your day at the show. Breed judging is subdivided by color group or variety, sex, and age. As you gain experience showing, you will become familiar with the order of classes. For most breeds, the varieties are judged in the order they are listed in the *Standard of Perfection.* Within each color group or variety, there may be different orders of judging in different parts of the country. One of the more common orders is as follows:

- Senior bucks
- Senior does
- Intermediate bucks
- Intermediate does
- Junior bucks
- Junior does

If your breed will be judged first or second, you will want to direct your efforts to final pre-show preparations. If your breed is toward the end of the list, you'll probably have time to look around, have a snack, and visit with other exhibitors. At most shows, rabbits remain in their carriers until it is their turn on the show table. Under the best circumstances, you can place your carrier in or near the judging area. You will want to be within hearing distance when judging for your rabbit's class is announced.

The Judging Process

When it is time for your class, carry your rabbit and its coop card to the judging area. Carefully place your rabbit into one of the divided holding sections on the judge's table. Once your rabbit is secure, take your coop card to the writers who are seated beside the judge's table. Your work is done for now, so find a good spot from which to watch the judging.

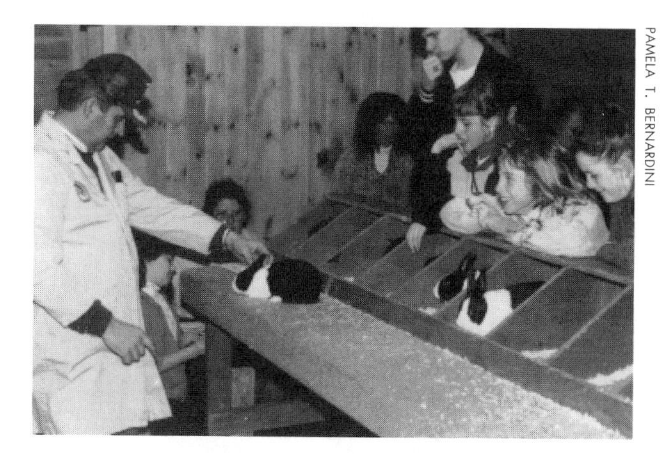

PAMELA T. BERNARDINI

The judge comments on the good and not-so-good points of each rabbit.

The judge will begin judging the class by checking each animal for eliminations or disqualifications. If an elimination or disqualification is found, that animal is out of the competition and is returned to its owner. Next, the judge will begin to compare the rabbits that remain on the table. You may see the judge handle your rabbit several times as he or she compares it to others in the class. The actual placement of the rabbits begins at the bottom of the class. As the judge ranks each rabbit, he or she will evaluate the animal and comment on its good and not-so-good points. The writers record these comments on the animal's coop card. As each rabbit is placed and the comments are recorded, the owner picks up the rabbit and returns it to its carrier. The last rabbit on the table is the winner of the class!

Class winners remain at the judging area until all classes in the variety or group have been judged. The judge then compares all the class winners and decides which animal is Best of Variety or Group. The judge also selects a Best Opposite. The Best Opposite is the animal that the judge believes is the closest to the Best of Variety or Group animal, but is of the other sex. Therefore, if the Best is a buck, the Best Opposite will be a doe; if the Best is a doe, the Best Opposite will be a buck.

PAMELA T. BERNARDINI

Winning a trophy for your rabbit is only one of the benefits of going to shows.

After all varieties or groups within a breed have been judged, the judge chooses Best of Breed and Best Opposite of Breed. To make this decision, the judge compares all the Best and Best Opposites of the color varieties. Selection of Best of Breed and Best Opposite of Breed completes the judging for the breed. At the end of the show, the Best of Breed animal will compete with the winners from the other breeds for the top honors of the day.

Awards and Prizes

After your rabbit has been judged, you may be anxious to find out what you have won. If your rabbit placed near the top of its class (usually first to fifth place), there will probably be a ribbon award. There may be a special rosette and perhaps a small cash prize if your rabbit was good enough to win Best of Variety or Group. The Best of Breed prize is often a trophy and a small cash award. Winners of top awards such as Best in Show usually receive a large trophy as well as prize money. Ribbons and trophies are usually presented at the show. Cash prizes are often mailed to winners after the show.

Another type of prize your rabbit can win is a *leg*. Legs are special certificates that are issued at A.R.B.A. shows. If your rabbit wins its class and there were at least five rabbits in the class being shown by at least three different exhibitors, then your rabbit will receive a leg. Legs are also issued for Best of Variety or Group, Best of Breed, and other prize categories. Legs are mailed to exhibitors after the show. After a rabbit receives three legs, its owner can apply to the American

AMERICAN RABBIT BREEDERS ASSOCIATION, Inc.
Record of Leg of Grand Champion Certificate

OWNER ____________________
ADDRESS ____________________
CITY ____________ STATE ________ ZIP ________

DATE ________ ARBA SHOW SANCTION NO. ________
SHOW ____________________
WHEN ________ WHERE HELD ________
SECRETARY ____________________
JUDGE ____________________

COLOR		BREED
CLASS	SEX	BORN
EAR NO.	REGISTRATION NO.	

First Place		No. In Class		No. breeders exhibiting in class	
Best of Breed		No. in Breed		No. breeders exhibiting in breed	
Best Opposite Sex of Breed		No. Opp. Sex of Breed		No. breeders exhibiting Opp. sex of breed	
Best of Group		No. In Group		No. Breeders exhibiting in group	
Best Opposite Sex of Group		No. Opp. Sex of Group		No. Breeders exhibiting Opp. sex of group	
Best of Variety		No. In Variety		No. breeders exhibiting in Variety	
Best Opposite Sex of Variety		No. Opp. Sex of Variety		No. breeders exhibiting Opp. sex of variety	
Best in Show		No. in Show		No. exhibitors in show	

RULES GOVERNING AWARDING LEGS ON GRAND CHAMPION

A Leg on Grand Champion will be awarded to any rabbit or (Cavy) that
1. Wins First in a class of not less than 5 entries owned by 3 or more exhibitors.
2. Wins Best of Breed providing there is 5 or more shown in the breed by 3 or more exhibitors.
3. Wins Best Opposite Sex of Breed providing there is 5 or more of the same sex as the winner shown in the breed by 3 or more exhibitors.
4. Wins Best of Group providing there are 5 or more shown in the group by 3 or more exhibitors.
5. Wins Best Opposite of Group provided there are 5 or more of the same sex as the winner shown in the group by 3 or more exhibitors.
6. Wins best of Variety providing there is 5 or more shown in the variety by 3 or more exhibitors.
7. Wins Best Opposite of Variety providing there is 5 or more of the same sex as the winner shown in the variety by 3 or more exhibitors.
8. Wins Best of Show.
9. Two Legs cannot be honored for the same show on the same animal.
10. At least one of the wins must be obtained as an intermediate or a senior and the wins must be under at least 2 different A.R.B.A. Judges.
11. No Leg can be awarded unless it is attained in an officially sanctioned A.R.B.A. show under an A.R.B.A. licensed judge.
12. The Leg for Grand Champion must be obtained by the exhibitor from the show secretary where the award was won.
13. Rabbits must be judged by a Licensed ARBA Rabbit Judge and Cavies judged by a Licensed ARBA Cavy Judge.

RULES GOVERNING AWARDING GRAND CHAMPION CERTIFICATE ON REVERSE SIDE.

REV 1/89

A.R.B.A. record of leg.

Rabbit Breeders Association to receive a Grand Champion certificate. Achieving a Grand Champion award is quite an accomplishment. It will also increase the value of your rabbit for sale and breeding.

Other Opportunities at Rabbit Shows

Although the judging of breeds is the major focus at a rabbit show, don't overlook the other show opportunities.

A unique learning opportunity. Watching the breed judging can teach you a lot, but don't overlook the other information sources that are present at shows. When you are in a show room, you are literally surrounded by hundreds of years of rabbit-raising experience. Take advantage of opportunities to talk with other exhibitors. Most exhibitors enjoy talking about their favorite subject—their rabbits. Sharing rabbit-raising experiences is a way we can learn from each other's successes and failures.

An opportunity to develop new skills. Many youth shows offer classes to help you sharpen your rabbit-related skills. Showmanship contests are probably the most common and the most popular. In a showmanship contest, you and your rabbit are a team. How well you handle your rabbit, how much you know about caring for your rabbit, and the condition of your rabbit are the important factors in a showmanship class.

Rabbit-judging contests are another form of youth competition. A judging contest gives you a chance to be the judge and place a class of four rabbits. How close you come to the placing made by an official judge is what will determine your score in the judging contest.

Other youth contests include breed identification and dress-up classes, as well as photography and arts and crafts that feature rabbits. If you attend a show that offers these competitions, join the fun and participate.

Opportunity to buy, sell, and trade. Part of the fun of attending a rabbit show is the opportunity to see so many different breeds for sale at one location. A show can be a real rabbit marketplace. If you choose to buy a rabbit at a show, be sure to take your time and thoroughly inspect the animal. Just because the animal is at a show does not necessarily mean that it is of show quality. If you sell rabbits at a show, offer quality animals that will be good representatives of your rabbitry.

Opportunity to examine and purchase rabbitry supplies. Rabbitry supply businesses often set up booths at rabbit shows. This comes in handy because many items that are useful in the rabbitry are not available at your local stores. Supplies purchased at a show can also save you money because you will not have to pay shipping charges.

A chance to support the sponsoring club. Rabbit shows don't just happen. They take lots of hard work, and this work is done by people who volunteer their time. The sponsoring club usually has a food booth and conducts a raffle to help with the costs of organizing a show. You support the show when you purchase a snack, buy a raffle ticket, or donate an item for the raffle. You also support the club when you donate your time. When you are not busy showing your rabbits, you may be able to help by selling raffle tickets, lending a hand at the food booth, or assisting writers with the paperwork.

After the Show

After the show, you may have a long ride home. Although it may be late when you arrive, remember that your rabbit has had a long day, too. Don't leave your rabbit in its carrier and figure that it will be okay until the morning. Take the time to get your animal settled back into its cage, with fresh food and water. If you have other rabbits and did not make arrangements for a family member to feed them, you will also have your regular evening chores awaiting you.

Showing rabbits can make for some pretty long days! But when I think about the positive comments the judge had for my rabbits, the new friends I made, or the old friends I have just seen again, somehow all the work seems worth it. I guess you have to be a rabbit person to understand.

Marketing Your Rabbits

As much as you might like to keep every rabbit that you raise, there will come a time when you decide to sell some of your stock. Selling rabbits is a natural and necessary part of a rabbit-breeding project. Will you make lots of money selling rabbits? Probably not, but you will bring in money to help pay for feed and other expenses. If you get really good at selling rabbits, your rabbit project can show a profit or grow into a part-time business.

One way to increase your potential for profits is to *diversify*. This means to offer a variety of products instead of just one. If you decide to sell rabbits only for pets, you are offering just one product. Think about how many more customers you might have if you also sold pedigreed breeding stock, had meat rabbits available, offered pet care for vacationing rabbit owners, and offered, for a fee, such services as tattooing, grooming, and nail-trimming.

Selling Rabbits for Pets

Selling your rabbits for pets can be a very large part of your marketing plan. More and more people are discovering what great pets rabbits can be. Since rabbits can be kept indoors or outside, they can fit into most households.

If you decide to sell pet rabbits, you quite likely will sell to many people who do not have experience taking care of rabbits. Part of your job as seller should include teaching the new owners about proper rabbit care. Be patient and spend time answering questions. You might want to have some written material to give to new rabbit owners. For example, the A.R.B.A. has some pamphlets on pet care that are free or very inexpensive. A little extra effort will mean a better home for the bunny you are selling, plus a satisfied customer who may send more business your way.

Selling Rabbits for Breeding Stock

When customers buy breeding stock, they are buying an animal plus the future animals they hope to produce. Animals sold for breeders should be the best that you have. They should be quality rabbits that will produce good examples of their breed. Not every animal in a litter will measure up. Since rabbits sold for breeding stock and for show are your best, these are the animals that will have the highest price tag.

With every rabbit you sell as breeding stock, provide a three-generation pedigree for the animal.

Selling Rabbits for Meat

Rabbit meat is a white meat that tastes good and is good for you. It is low in cholesterol and fat, and it is highly digestible. Having meat rabbits available is another way to attract more customers.

Young rabbits that weigh about 4 to 5 pounds are called fryers, and many people prefer this size. The meat of older, larger rabbits is usually not as tender as the meat of fryers, so you may have to sell these at a lower price per pound.

Meat rabbits are sold *live* or *dressed*. Dressed rabbits sell for more money because they are ready to cook. Live rabbits cost less per pound because the buyer has to arrange for the processing of the animal.

Attracting Customers

Once you have some rabbits to sell, the next step is to find buyers for them. Advertising your stock doesn't have to cost a lot. There are several no-cost and low-cost methods to get the word out.

Talk with Friends

Let your friends know that you have rabbits for sale. Friends that enjoy visiting your rabbits may be considering getting a rabbit of their own.

Give Free Demonstrations

Day care centers, schools, and Granges are just a few of the places where there are groups that might enjoy learning about rabbits. If you volunteer to talk to a group about rabbits, you might find that some of the members become future customers.

Take Advantage of Free or Low-Cost Advertising Opportunities

Bulletin boards in stores. Today, many stores have bulletin boards where customers may place free ads. Take a little time to make an attractive card or poster to advertise your rabbits. Using color and pictures will help get the attention of potential buyers.

"Rabbits for Sale" sign at your home. Your sign

What Your Ad Should Say

- *What you have.* Be specific. "Bunnies for Sale" doesn't tell as much as "Adorable Holland Lops! Show Quality, Pets & Breeding Stock Available."
- *Whom to contact.* List all the information that will make it easy for people to reach you: name, address, and telephone number.
- *Features that make your rabbits special.* For example: "All registered stock," "Free booklet on rabbit care."

should be neat, attractive, and easy to read from the road. Put it out when you have rabbits available, and take it down when you do not have stock for sale.

A colorful poster will catch a potential buyer's eye.

Make a neat, easy-to-ead sign to advertise at your home.

Shows. If you show your animals, the show room can be a good place to advertise. Lots of people will see your rabbits at a show. Some shows have special sale areas. Some have special rules about selling animals at the show. Be sure to check the show catalog for any rules that affect the sale of your animals.

Business cards. Purchasing business cards from a printer or stationery store is often a good investment. They range in price but are usually fairly inexpensive. Business cards give your rabbitry a professional image, and cards are easy to post or hand out at shows. If you create a business card for your rabbitry it should include

- The name of your rabbitry
- The breeds and services that you offer
- Your name, address, and phone number

Create your own business card.

Customer Relations

No matter what type of business you are in, the best form of advertising is satisfied customers. If people are happy with your rabbits, they will tell their friends. If people are unhappy with your rabbits, they will also tell their friends. Your rabbit business will do much better if customers have good things to say about you and your products and services. Treat your customers the way you would like to be treated. To establish and maintain good customer relations, you should

- Respond to questions about stock. If people call or write, let them know what you have for sale. If you don't have stock available, let them know when you will. Being ignored turns customers off.
- Sell only healthy animals. Take time to check each animal before it leaves your rabbitry. New rabbit owners may not know how to choose a healthy animal, and they rely upon your ability to do so.
- Be sure of what customers want. Use your best judgment to sell them animals that meet their needs. If a customer wants a small apartment pet, don't offer that cute little 5-week-old Checker

Giant. If you sell pet-quality rabbits, explain to customers why the animals they are buying from you are not suitable for show.

- Do all you can to help new owners understand good rabbit care. Answer their questions, and provide some written information for them to take home.
- If something goes wrong, do your best to make it right. Replacing or exchanging an animal may not be convenient, but the goodwill it buys is more valuable than the cost of another rabbit.

New Rabbit Owner's Kit

This kit is simple to put together, and it can be very helpful to new owners. You may want to have a few made up at all times. What you need for each kit:

- 1 food storage bag (10 x 13 inches)
- 1 sandwich bag
- rabbit pellets
- fact sheet or brochure on rabbit care
- business card or tag with your name and telephone number

Prepare a new rabbit owner's kit containing feed and rabbit care information.

Fill the sandwich bag with pellets and seal it so the pellets won't fall out. Many new owners have not thought to buy food for their new rabbit, and this small bag of feed will ensure that the rabbit is properly fed in case the new owners cannot purchase feed until the next day. If the new owner will be using a different feed, having the feed the rabbit is used to allows a gradual change in diet.

Place the bag of pellets and the written information about rabbit care into the larger bag. Tie the top of the bag and attach a business card or tag.

CHAPTER 10

Managing Your Rabbitry

All the skills and knowledge that you have gained will benefit your rabbits only if you apply them through a sound rabbitry management program. The dictionary defines management as "the art of handling, controlling, and directing." As the manager of your rabbitry, you are responsible for its success. This can seem like a pretty big job, but big jobs become a lot easier when you break them down into small jobs. A good management program will allow you to use your time effectively to accomplish the many tasks that are part of being a successful manager. The tasks that are required for good rabbitry management can be broken down into

- Daily tasks
- Weekly tasks
- Monthly tasks
- Seasonal concerns
- Tattooing
- Record-keeping

Things to Do Daily

Feeding and watering are basic. Establish a daily schedule and stick with it, 7 days a week.

Observe your rabbits and their environment. Daily observation helps you catch small problems before they become large problems.

Keep things clean. Attend to small cleaning needs so they don't grow into large cleaning chores.

Handle your rabbits. Regular handling will make your animals more gentle, and you will become more aware of their individual condition. This may be impractical in a large rabbitry, but you should strive to handle some of your animals each day.

Things to Do Weekly

Clean cages. Solid-bottom cages and cages with pull-out trays must be cleaned and re-bedded weekly. On wire-bottom cages, use a wire brush to remove any build-up of manure or fur.

Clean feeders. Rinse crocks with a water-and-chlorine-bleach solution (1 part household bleach to 5 parts water). Check self-feeders for clogs of spoiled feed.

Check rabbits' health. If there are animals that you did not have time to handle and check earlier in the week, take time and do so now.

Check supplies. Do you have enough feed and bedding for the coming week? Your family will appreciate knowing ahead of time if a trip to the grain store is going to be needed.

Make necessary repairs. Have you noticed a loose door latch or a small hole in the wire flooring? Take time to do these small repairs before they lead to larger problems.

Prepare for coming events. Is a doe due to kindle in the next week? Is a show entry due soon? Check your rabbitry calendar, where these things should be noted. Do what's necessary to be ready.

Check growing litters. Is the nest box clean? Is it time to remove the nest box? Is there any evidence of eye infections? These are just a few of the conditions to check in developing litters.

Things to Do Monthly

Check toenails. You will not have to trim the toenails of every rabbit every month, but you should check each animal and trim those that need it. This is an important management skill to learn, because properly trimmed toenails decrease the chances of your rabbit's being injured. Long toenails can get caught in the cage wire and cause broken toes and missing toenails. The time spent trimming toenails will also benefit you. If your rabbit's toenails are properly trimmed, you will be less likely to be scratched when you handle your rabbit.

Trimming toenails is not hard to do, but many rabbit owners put this task off because they are afraid of hurting their rabbit. Learning to trim toenails is easier if you have the help of a friend—one person can hold the rabbit, while the other does the trimming. The kind of trimmers you use on your own fingernails and toenails work well on most rabbits. Ask your parents if there is an extra pair you can use for your rabbits. Larger breeds sometimes need a larger trimming tool; dog toenail clippers work well for these breeds.

The person holding the rabbit should sit down with the rabbit supported on his or her legs and turned over for trimming. The person who is doing the trimming can now use both hands—one to push the fur back in order to see the nail, and the other hand to do the trimming.

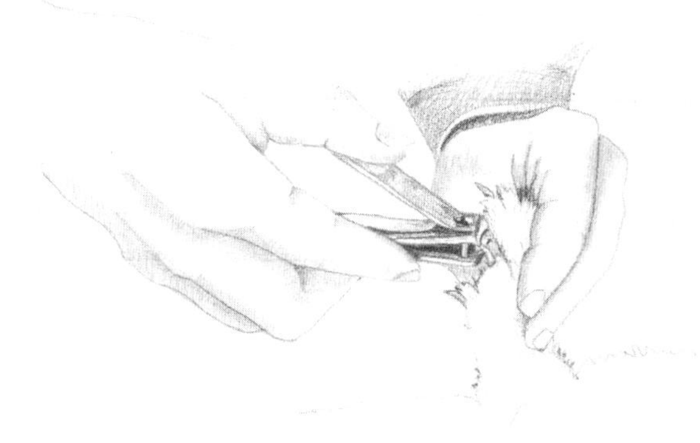

Trim the white section of your rabbit's toenails; if you trim the pink section, you will hurt your rabbit.

If you can, choose a white rabbit for your first trimming job. If you look at a white toenail, you will be able to see a pinkish line extending from the toe. This is the blood vessel that brings nourishment to the growing nail. Toenails that need trimming have a clear white section after the pink. This white area can be trimmed without hurting the rabbit. If you trim the pink area, it will hurt the rabbit. Leave about ¼ inch of nail beyond the end of the blood vessel. On colored toenails, it is more difficult to see where the blood supply ends. If you are not sure, trim only a small amount at a time. If you happen to cut too closely and the rabbit begins to bleed, use a clean cloth to apply direct pressure and stop the bleeding.

Update written records. Catch up on writing pedigrees. Record feed costs and other rabbitry expenses, so you will have a sense of how much your rabbit project is costing.

Provide preventative medicines. If you have identified a need to offer a preventative medicine, most (such as preventatives for coccidiosis) are offered on a monthly schedule. Your preventative program will be more effective if you administer it on a regular schedule.

Tend to the needs of developing litters. Young rabbits grow a lot in one month's time. Litters should be weaned by 8 weeks of age. This is also the time to tattoo and to separate littermates by sex.

Check fans and air vents, if your rabbitry is indoors. Good ventilation is extremely important to the health of your rabbits.

Seasonal Concerns

Climates vary from one part of the country to another, but most sections of our country do experience a variety of weather conditions. You or I can put on a jacket or turn on the heat when it's cold, and we can go swimming or turn on the air conditioning when it's hot. Although we can't provide these same luxuries for our rabbits, there are things we can do to manage their environment and make them more comfortable.

Cold-Weather Care

Rabbits do quite well in cold weather and can survive temperatures well below zero. However, you do need to provide protection from winds, rain, and snow. If your cages are inside, you are already providing this protection. If your cages are outside, you will want to add protection as the temperature drops. Most outdoor cages can be enclosed quite easily by stapling plastic sheeting around three sides. If you live where the weather is extremely cold, cover the front with a flap of plastic as well. Do not enclose the whole cage so tightly that it is difficult to get in to feed and care for your animal or that the ventilation is poor. If your feed comes in plastic grain bags, the empty bags can be used for covering material. (See page 28 for illustration.)

Try to take advantage of the winter sun. If your outside cages are movable, place them in a location that receives a lot of direct sun.

Although healthy adult rabbits don't suffer when it's cold, newborn rabbits can easily die from the cold. If you are expecting a litter during the cold weather, be sure the doe has a well-bedded nest box. You may want to move the doe into a warmer location, such as a cellar or garage. After the litter is about 10 days old, the cage, with Mom and her nest box of babies, can be moved back outside. Some breeders keep the nest box of newborn rabbits in their warm homes. At feeding times, they take the box out to Mom or bring her inside to nurse her litter.

Hot-Weather Care

Special care is more important during hot weather than during cold. Fur coats that keep rabbits cozy in the winter can sometimes provide too much warmth during the summer months. Here are some rules to keep your rabbits comfortable during a heat wave:

- Place outside hutches in shady locations.
- Provide enough ventilation. Remove plastic sheeting or hutch cover. If your hutches are inside, open windows or provide a fan to increase air circulation.
- Provide lots of cool, fresh water.
- Be prepared for extreme heat. Use empty plastic soda bottles to make rabbit coolers. Fill the bottles two-thirds to three-quarters full with water, and keep them in your freezer. In periods of extreme heat, lay a frozen bottle in each cage. The rabbits

will beat the heat when they stretch out alongside their rabbit cooler. It's also a good idea to place a cooler in the carrier if you are traveling with a rabbit in the hot weather.

Tattooing

Tattooing provides a permanent method to tell who's who in your rabbitry. Tattooing is important for showing, but it is equally important for good record-keeping, and should be a regular part of your rabbitry management program. If you have a small rabbitry, it may be impractical to spend money for tattooing supplies. Ask other breeders if they will tattoo your animals. Chances are that 4-H leaders and older club members will be willing to help you tattoo your rabbits.

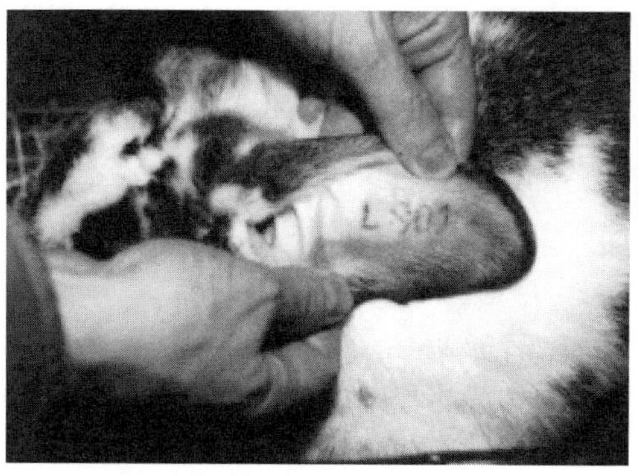

PAMELA T. BERNARDINI

Ear tattoos clearly identify individual rabbits.

Your Tattoo System

The tattoo can be a number, a word, or a combination of letters and numbers. Some breeders use short names like Joe or Pat, so that the tattoo names the animal as well as identifies it. Others develop systems that tell more information about each animal. For example, in one system, a letter stands for a certain buck, another letter stands for a certain doe, an odd number means the animal is a buck and an even number means it is a doe. Thus, if we had an animal with the tattoo "TB3," we would know that it was a buck whose father was named Thumper and whose mother was named Bambi. The possibilities go on and on, and you can develop a tattoo system that works best for you. Most tattoo sets are limited to five numbers or letters per tattoo, so be sure your system takes this limitation into account.

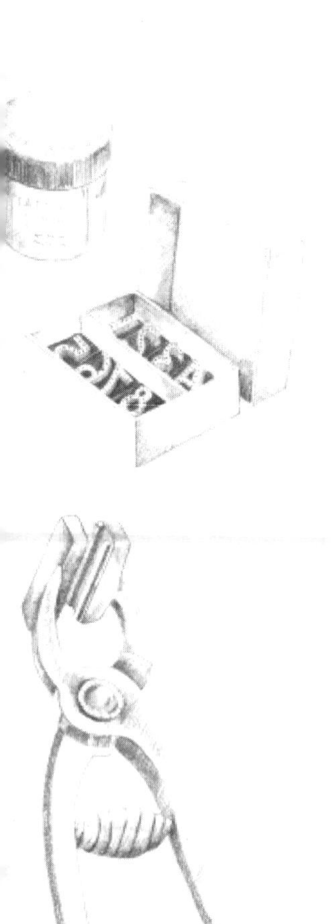

A tattoo set for permanently marking rabbits' ears

What Is a Tattoo?

If you look at the tattoo set, you will see that the letters and numbers are made up of a series of small needle-like projections. When the tattoo pliers are closed on the rabbit's ear, they make a series of small punctures. Tattoo ink is then rubbed into the area. The tattoo ink that goes into the punctures becomes permanent, while the ink on the ear's surface soon wears away.

You may be wondering if tattooing is painful for the rabbit. The answer is yes, but only for a very short time.

Learning to Tattoo

As your rabbit project grows, you will want to learn how to tattoo your own animals. Observe others when they tattoo. Everyone develops his or her own technique. There is no one right way, but here are some tattooing tips you might find helpful.

- Practice placing the numbers and letters into the tattoo pliers. Use a piece of paper to check what you have in the pliers. It's very easy to have a 21 when you really wanted a 12. By checking on paper first, you'll avoid an incorrect tattoo.
- Choose rabbits that are not show quality for your first tattoos.

- Use a little rubbing alcohol on a cotton ball to wipe out the ear before tattooing. This cleans the ear surface and allows the ink to adhere better.
- Before you tattoo, look at the inside of the rabbit's *left* ear. You will be able to see some of the blood vessels that are just below the surface of the skin. Find a space where no large vessels cross and plan to place your tattoo there. Even with careful planning, you may hit a blood vessel. This will cause some bleeding, but it can be easily stopped by applying some gentle pressure to the area. Bleeding can wash the tattoo ink out of the punctures, so avoiding blood vessels usually makes for a more successful tattoo.

Get a friend to completely restrain your rabbit while you clamp the tattoo pliers on the rabbit's ear.

- Restraining the rabbit during tattooing is very important. The rabbit will pull away when the pliers are clamped, and the rabbit could be injured. If you have someone to help you, one person can restrain the animal and the other can do the tattooing. The holder can wrap the rabbit in a towel so only its head and ears stick out. The towel will keep the rabbit from moving and from scratching the holder. When I tattoo alone, I place the rabbit in a narrow show carrier. This keeps the rabbit confined and allows me to use both hands for the tattooing. One hand secures the left ear and the other holds the tattoo pliers.

- Clamping the pliers on the rabbit's ear is the hardest part, because no one wants to hurt an animal. Try to take comfort in the fact that the pain lasts a very brief time. Release your grip on the pliers immediately after you close them, and the worst is over. It is difficult to say exactly how hard to close the pliers to ensure a good tattoo. If you are tattooing young rabbits, their thin ears will need only the pressure used to pre-test the tattoo on paper. Older animals have thicker ears, and will need a slightly firmer grip.

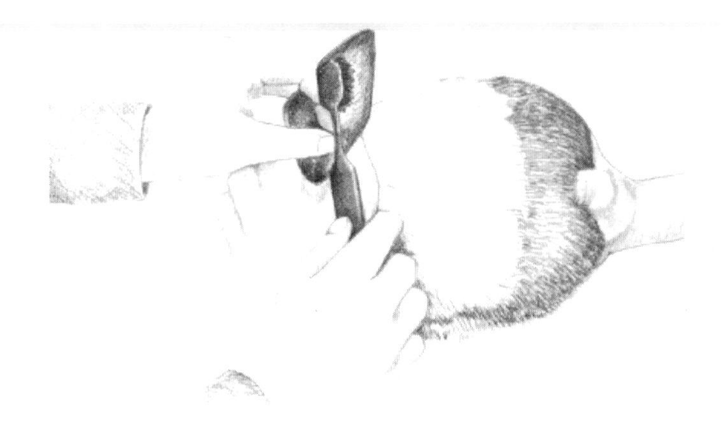

Apply ink to the tattoo with an old toothbrush.

- Give the rabbit a few seconds to calm down, and then look inside the ear. Your tattoo should appear as a series of dots. Now you are ready to apply the tattoo ink. I like to use an old toothbrush for this.

A toothbrush is a good size, has a convenient handle, and the bristles help to work the ink into the small punctures. Most rabbits do not seem to mind this part of the procedure.

- After you've applied the ink, apply a thin layer of petroleum jelly to the tattoo area. This forms a kind of seal that helps to keep the ink in the punctures. The excess ink will wear off normally or the mother may clean it by grooming her baby. If the ink has not worn away by the time your rabbit is entered in a show, you will need to remove it to make the tattoo readable. Use a tissue and a little petroleum jelly to wipe the excess ink away.

Keeping Records

If you own one or two rabbits, it will be easy to remember all the important details about them. As your rabbits begin to produce litters and the size of your rabbitry grows, however, it will be difficult to recall everything about each animal. Keeping accurate written information about your rabbits is important. This paperwork can help in many ways to increase the success of your rabbit project.

- You will have information to help you decide which doe gets bred to which buck.
- Your rabbitry will be more productive because your records will show which animals are your top producers. You will use this information to help you decide which animals to keep in your herd.
- You will know how much your rabbits cost you and how much money they make for you.
- You will have the information you need to make a written pedigree for each animal. A rabbit with a pedigree usually brings a higher price.

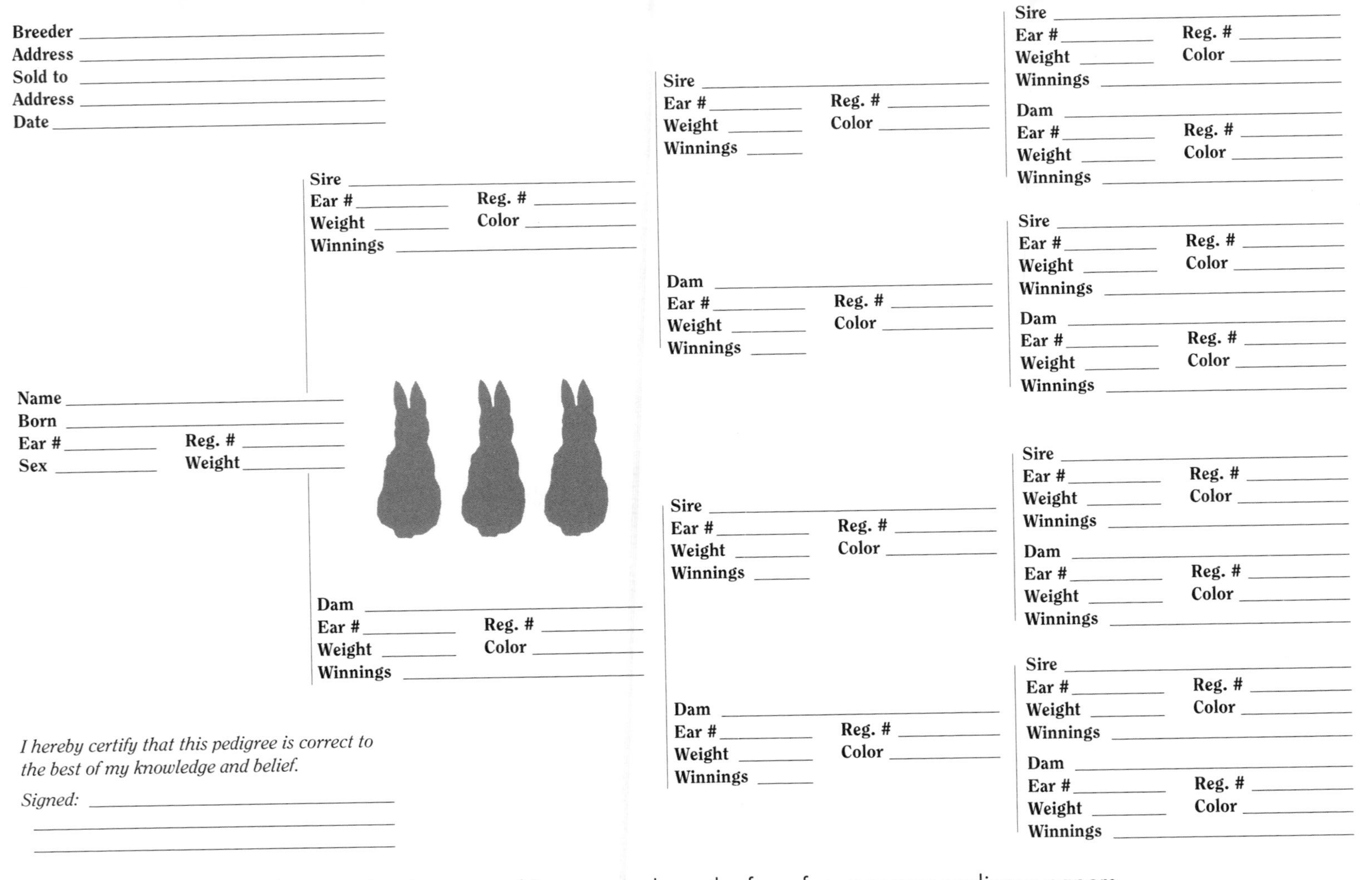

Breeder ______
Address ______
Sold to ______
Address ______
Date ______

Name ______
Born ______
Ear # ______ Reg. # ______
Sex ______ Weight ______

I hereby certify that this pedigree is correct to the best of my knowledge and belief.

Signed: ______

Sire ______
Ear # ______ Reg. # ______
Weight ______ Color ______
Winnings ______

Dam ______
Ear # ______ Reg. # ______
Weight ______ Color ______
Winnings ______

Sire ______
Ear # ______ Reg. # ______
Weight ______ Color ______
Winnings ______

Dam ______
Ear # ______ Reg. # ______
Weight ______ Color ______
Winnings ______

Sire ______
Ear # ______ Reg. # ______
Weight ______ Color ______
Winnings ______

Dam ______
Ear # ______ Reg. # ______
Weight ______ Color ______
Winnings ______

Sire ______
Ear # ______ Reg. # ______
Weight ______ Color ______
Winnings ______

Dam ______
Ear # ______ Reg. # ______
Weight ______ Color ______
Winnings ______

Sire ______
Ear # ______ Reg. # ______
Weight ______ Color ______
Winnings ______

Dam ______
Ear # ______ Reg. # ______
Weight ______ Color ______
Winnings ______

Sire ______
Ear # ______ Reg. # ______
Weight ______ Color ______
Winnings ______

Dam ______
Ear # ______ Reg. # ______
Weight ______ Color ______
Winnings ______

Sire ______
Ear # ______ Reg. # ______
Weight ______ Color ______
Winnings ______

Dam ______
Ear # ______ Reg. # ______
Weight ______ Color ______
Winnings ______

If you wish, photocopy this page and use the form for your own pedigree papers.

Pedigree Papers

A rabbit's pedigree papers are a written record telling who its parents, grandparents, and great-grandparents were. A pedigree is the rabbit's family tree. Usually, pedigrees also include information such as the color, tattoo number, and weight of each animal listed. You can use the information on a pedigree to help you make decisions that will affect your breeding program. Are you trying to increase the weight of your rabbits? If you breed your doe to a buck whose pedigree shows relatives with high weights, you will have a better chance of producing heavier rabbits than if you breed to a buck whose relatives are on the light side. Are you trying to produce a certain color of Mini Lops? Even if your buck and doe are not of that color, if the color appears in their pedigrees, it is more likely to appear in their offspring.

As the breeder, it is your responsibility to keep accurate pedigrees for the rabbits that you raise. Most breeders make out pedigrees for the animals used for showing and breeding purposes. Pedigree papers are not necessary for animals sold for pets or meat.

Filling out pedigrees can be confusing at first. You have to take information from the buck's pedigree and combine it with information from the doe's pedigree. Pedigree forms use the term *sire* and *dam*. Sire refers to the father, or the buck. Dam refers to the mother, or the doe. A fun way to practice filling out pedigrees is to make one out for yourself or for a person you know. If you make your own pedigree, your name will appear instead of the rabbit's name. The sire will be your dad. Information about his parents and grandparents (your grandparents and great-grandparents) will complete the top half of the pedigree. You will list your mom where the dam's name goes and complete the bottom half of the pedigree with information about her parents and grandparents.

When you sell a purebred rabbit, you will want to

Sources of Blank Pedigrees

- The American Rabbit Breeders Association sells pedigrees in books of 50.
- Most breed associations have pedigrees that feature an illustration of their breed.
- You can create your own pedigree form.
- You can duplicate the pedigree found on page 128 of this book.

Sire. *Father, or buck.*

Dam. *Mother, or doe.*

have a pedigree for the buyer. It takes time to fill out a pedigree. If you have it written up ahead of time, you won't have to mail it to the buyer later.

Registration Papers

Many people mistakenly think that pedigreed animals are the same as registered animals. A registered rabbit must have a pedigree, but it has to meet additional requirements as well.

- The rabbit must be examined by a registrar who is licensed by the American Rabbit Breeders Associa-

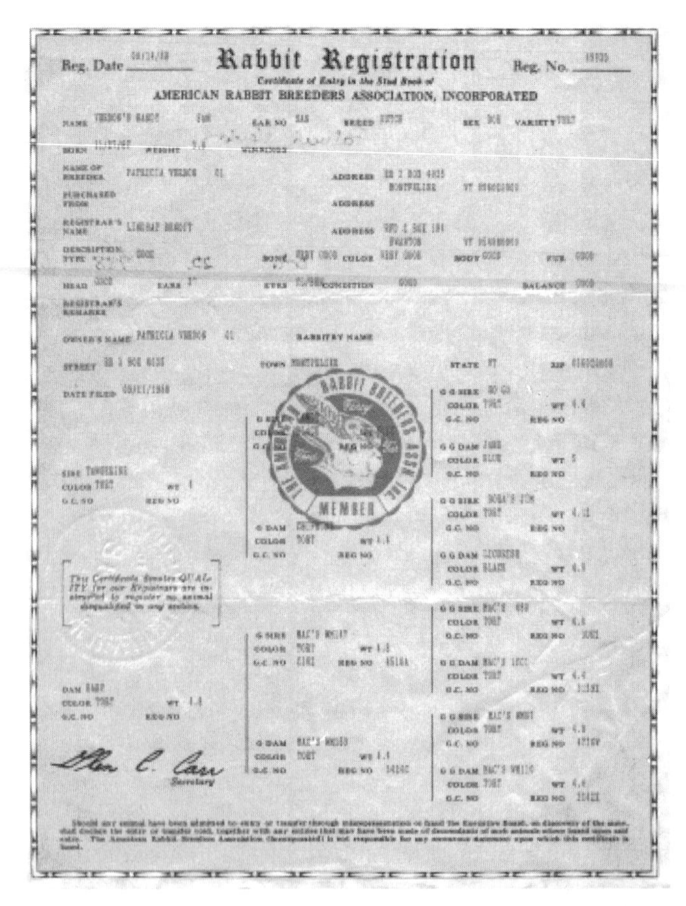
Reg. Date ______ **Rabbit Registration** Reg. No. ______

Certificate of Entry in the Stud Book of

AMERICAN RABBIT BREEDERS ASSOCIATION, INCORPORATED

NAME — EAR NO — BREED — SEX — VARIETY

BORN — WEIGHT — WINNINGS

NAME OF BREEDER — ADDRESS

PURCHASED FROM — ADDRESS

REGISTRAR'S NAME — ADDRESS

DESCRIPTION: TYPE — BONE — COLOR — BODY — FUR

HEAD — EARS — EYES — CONDITION — BALANCE

REGISTRAR'S REMARKS

OWNER'S NAME — RABBITRY NAME

STREET — TOWN — STATE — ZIP

DATE FILED

Secretary

Sample A.R.B.A. registration form.

tion. The registrar will come to your home and check to see that your rabbit has no eliminations or disqualifications for its breed. The registrar will need a copy of the rabbit's pedigree.

- You will have to show that you are a member of the A.R.B.A.
- You will have to pay a registration fee (about $4).

If your rabbit passes its examination, is at least 6 months old, and has a permanent tattoo number in its left ear, then it can be registered. The registrar will fill out papers to send to the A.R.B.A. and will tattoo a registration number in your rabbit's right ear. Your rabbit's official registration papers will be mailed to you from the A.R.B.A.

If you decide to register your purebred rabbits, plan ahead and have several done at once. You may want to get together with other rabbit owners and plan a time for a registrar to come. You can also have your rabbits registered at A.R.B.A. shows, which have a registrar on duty. A rabbit show is a convenient place to have your animals registered. Just remember to bring each rabbit's pedigree.

Hutch Cards

A pedigree tells who your rabbit is. A hutch card, on the other hand, tells what he or she can do. Hutch cards, as their name suggests, are attached to the rabbits' cages. They are kept handy so you can write down important information as it happens. For a doe, a hutch card records such information as which buck she is bred to, when her litter is due, and how many young she kindles. For a buck, the card lists which does he is bred to, when they kindle, and how many young he has fathered. Hutch cards help you see how productive your breeders are.

Hutch cards are often available free from feed

companies. You may be able to pick them up at your feed store, or you may have to write to the feed company and request a supply.

The information recorded on the hutch cards is important. Since they can be easily soiled or lost, it is smart to record the information in a second location. On a regular basis—perhaps once a month—transfer your hutch card information to a permanent management record.

Management Records

Management records usually cover a full year. They bring lots of information together so you can get a good overall picture of your total rabbit project. Using your management records, you will be able to keep track of things like the cost of feeds used, the total number of young born, the total income from rabbits sold, and your show results. If you keep good management records, you will know if your rabbits are productive, and you will know how much it costs to have rabbits as a hobby. Your management records will help you to decide if your hobby can become a profitable small business.

Many youth groups encourage record-keeping. You can write to 4-H and FFA to request their management record forms. If you become a member of one of these groups, your management records could win a prize.

Keeping records will be easier if you use these hints:

- Make record-keeping part of your regular routine. Do a little each week instead of trying to do it all at once.
- Make it easy to record information by keeping a pen or pencil handy.
- Have a shoe box or large envelope in a convenient spot for keeping receipts and show reports togeth-

er. Then, when you sit down to update your management records, you won't have to hunt all over for the information.

- Use a binder to keep pedigrees and management records safe and organized.

Post in a location that will be convenient to record the facts about your rabbitry. Monthly sheets can be totaled to determine an "annual report" of your rabbit project.

Expenses

Date	Animals Purchased		Pounds of Feed Used	Cost of Feed	Other Costs: Litter, Supplies, etc.	
	No.	Cost			Item	Cost
Total						

Income

(List of income from sale of equipment, breeding fees, etc.)

Date	Item	Amount Received
	Total	

Breeding Record

Date Bred	Name or Numbers		Date Due to Kindle	Date Kindled	Number of Live Young Born
	Doe	Buck			
				Total	

Show Participation

Date	Name and Place of Show	Placing or Award Received	Entry Fees	Value of Premiums Won
		Total		

CHAPTER 11

Activities for Young Rabbit Owners

Keeping your rabbits happy and healthy will take a lot of effort on your part. From time to time you may not feel very enthusiastic about all the chores that go along with being a rabbit owner. No matter how much you love and enjoy your animals, there will be days when other things seem like more fun. These feelings come to all rabbit people, young and old. Although adding more activities may not seem like a logical way to overcome the down times, I am absolutely convinced that getting involved in youth rabbit programs will help to keep you interested, challenged, informed, and motivated.

4-H

Although many people still think of 4-H as a program for farm kids, the 4-H program is conducted in every county in the country and draws members from all communities. 4-H Clubs give you an opportunity to share your interest in a subject—in this case, rabbits—

while enjoying activities with other kids.

What sort of things do 4-H members do? It would be hard to list them all, and different clubs plan programs that meet the interests of their members. However, here are some of the activities 4-H groups include in their club programs:

Club meetings. Groups meet on a regular basis determined by the club. Members serve as club officers and help run club activities. Meetings are a combination of learning activities and fun.

Competition. Many 4-H rabbit clubs organize and conduct their own youth rabbit shows. Other competitions that take place throughout the year include 4-H fairs, poster contests, record-book competitions, and photography contests.

Demonstrations. Club members practice presentations in which they instruct others about some phase of rabbit care. These demonstrations are presented to school groups and community organizations, or they may be part of a 4-H public-speaking competition.

Exhibits. Clubs create educational displays that teach others about rabbits. These are exhibited at such places as fairs, schools, and shopping malls.

Rabbit Quiz Bowls. This is a team event in which members respond to questions about rabbits. Quiz Bowls range from fun events at club meetings to competitions for state championships.

Special workshops. These training sessions are offered at the county and state levels, and include 4-H'ers from throughout the state. Workshops may be just about rabbits, or may include other subjects.

Leadership experience. Many 4-H activities provide

opportunities for you to develop and use a variety of leadership skills. There are committees to serve on and programs to help plan and conduct. In 4-H, members helping other members is an important focus.

Community service. Helping to make life better for others is also part of 4-H Club work. Members and their rabbits often visit nursing homes and schools. Members also participate in food collections, environmental projects, and other worthwhile efforts.

Camping. 4-H summer camps are very popular. Camp programs include a variety of activities such as swimming, crafts, and sports. Some camps have special themes, such as animal science, horseback riding, or shooting sports.

The above listing does not mention the best part of 4-H—the people. You will meet some of the nicest people, both kids and adults. These new friends will add a lot to the fun you can have with rabbits.

If you would like to find out more about becoming a 4-H member, look in your telephone book for the 4-H office that serves your county. You may find the office listed under 4-H Clubs, under Cooperative Extension, or under the name of your county. If you don't find it listed, contact your state 4-H office, which will direct you to a local office. State 4-H offices are located at the state land grant university in every state. A listing of these state offices is included on pages 141–43 of this book. 4-H is free, and is open to all kids 6 to 19 years old.

American Rabbit Breeders Association

Lots of 4-H members also become youth members of the American Rabbit Breeders Association. Beyond the books and pamphlets, the shows, and the rabbit registration program, the A.R.B.A. also offers a variety of activities just for their youth members. Youth programs are a large part of the A.R.B.A. National Convention and Show. This event is held each year in a different city. Members can participate in many of the youth activities even if they are unable to travel to the convention. Some of the A.R.B.A. activities that youth members from all over North America can be part of include these:

National Youth Rabbit Show. This is the big one, in which thousands of rabbits raised by kids like you compete. Your rabbit can be shipped to the show or a friend can take it if you are unable to attend.

Management contest. Here is a chance to see how your rabbitry stacks up against the rabbitries of other youth members. You do not have to attend the convention to participate in this contest. Management winners are chosen based on a written questionnaire they complete. This contest is subdivided by your age, where you live, and the size of your rabbitry. The management contest looks at your rabbitry's production, housing, animal health, record-keeping system, and feeding program.

Achievement contest. Like the management contest, the achievement contest is based on a written form, so you don't have to be at the convention to take part. This competition looks at what you have accomplished with your rabbit project. Sharing your project and helping others learn about rabbits are important factors, along with production and show results.

Rabbit-judging contest. This is a hands-on contest, so you must be present to participate. With other kids your age, you will place four classes of rabbits. Each class consists of four rabbits, and you will have 7 minutes to decide which should be first, second, third, and fourth.

Educational contests. Posters, drawing and painting, crafts, educational games, woodworking, and photography are among the categories that make up this competition. If you enjoy these activities, you should exhibit your work at this national contest. Of course, entries must be about rabbits. This is another contest that you can enter even if you cannot be there. If you do not know anybody who is attending, you can send your exhibits by mail.

Royalty contest. If you are up for a challenge, the royalty contest is for you. You will compete with boys or girls your own age. Since most phases of the contest are held at the annual convention, you must attend to be considered a candidate. Prior to the convention, you must complete an application on which you tell about your rabbit project. At the convention you will take a written test and participate in a breed identification contest. Each candidate also takes part in a special interview. The final part of the contest for candidates over 9 years of age is a judging contest. Candidates who are under 9 are scored on their ability to handle their rabbits. Each candidate receives a numerical score for each portion of the contest. Scores are totaled and the candidate with the highest score in his or her age category becomes the reigning monarch. All royalty contest candidates are honored at a special awards banquet.

If you become a youth member of the A.R.B.A., be sure to look into participating in these contests. Complete rules and information on how to enter are found in the A.R.B.A. Yearbook, which all members receive.

A.R.B.A. membership is a real bargain; annual dues are $5. You can become a member by writing to

American Rabbit Breeders Association, Inc.
8 Westport Court
Bloomington, IL 61704
309-664-7500
www.arba.net

Other Opportunities

Although 4-H and the A.R.B.A. offer many in-depth activities for rabbit owners, don't overlook the offerings of other youth organizations. If you are involved in Scouting, your rabbit project provides an opportunity to earn merit badges in rabbit production or pet care. Perhaps you can use your project to teach other Scouts and help them to qualify for a merit badge.

Your interest in rabbits can also be helpful in school. Many aspects of raising rabbits can be adapted to special projects such as science fairs, research papers, and oral reports.

Helpful Sources

Cooperative Extension Service

For more information on local education and youth programs in your area, contact the nearest Cooperative Extension office, located through

www.csrees.usda.gov/Extension

Notes

Notes

Glossary

abscess (n.). Collection of pus caused by infection.

Angora (adj.). Rabbit breed with long, soft wool.

antibiotic (n., adj.). Medicine that kills germs that cause disease.

breed (n., v.). Group of rabbits or other animals that share characteristics such as color, size, and fur type; produce offspring; raise.

broken (adj.). Fur color of rabbits that includes white and one other color.

buck (n.). Male rabbit.

characteristic (n.). Distinguishing feature; trait.

class (n.). Group of rabbits of the same breed, variety, age, and sex that are judged together in a show.

coccidiosis (n.). Disease of rabbits spread by droppings and soiled feed and bedding; symptoms include soft droppings, rough-looking coat, slow growth, and pot belly.

conditioning (n.). Grooming, feeding, and other quality care to ensure a rabbit is in top shape for a show.

coop card (n.). Card containing information that identifies a rabbit during a show.

crossbred (adj.). Born of parents who are of different breeds.

culling (n.). Process of separating rabbits according to intended use—e.g., show, breeding, pet, meat.

Cuterebra fly (n.). Insect that lays its eggs on rabbits; the larval stage lives as a parasite under the rabbits' skin.

dewlap (n.). Fold of loose skin under the chin of some female rabbits.

disinfectant (n.). Germ-killing cleaning substance (such as a chlorine bleach solution used for disinfecting cages or a medicinal solution used for cleaning wounds).

disqualification (n.). Permanent defect that means a rabbit can never be used in show competition.

doe (n). Female rabbit.

domesticated (adj.). Trained or adapted to being among humans and being used by humans.

dressed (adj.). Skinned and prepared for cooking.

ear canker (n.). Scabby condition in rabbits' ears caused by ear mites.

elimination (n.). Temporary defect that eliminates a rabbit from show competition.

fostering (n.). Having a mother rabbit accept as her own a baby rabbit that is not hers.

genitals (n.). Reproductive organs, especially those on the outside of the body.

group (n.). A system of subdividing a breed by having animals with similar color patterns placed together.

hare (n.). Mammal of the order Lagomorpha, similar to the rabbit but larger, with longer legs and ears.

heat stroke (n.). Illness caused by being exposed to extremely high temperatures.

hock (n.). First joint of the hind leg of a rabbit or other animal.

hutch (n.). Rabbit house.

hutch card (n.). Card kept on or near the hutch that identifies a rabbit and contains other important information, such as breeding program.

infected (adj.). Contaminated by germs, causing pain, redness, and swelling.

inherited (adj.). Passed on by a parent or other ancestor, as a characteristic.

instinct (n.). Animal's natural tendency to behave in a certain way.

kindle (v.). Give birth (special term used for rabbits).

lagomorph (n.). A mammal of the order Lagomorpha, which includes rabbits, hares, and pikas.

larval stage (n.). Wormlike form of a newly born insect.

leg (n.). In rabbit competition, certificate earned by winning a class at an A.R.B.A.-sanctioned show.

litter (n.). Baby rabbits or other animals born in one birth.

loin (n.). The part of an animal's body between the ribs and the hips.

lop-eared (adj.). Having bent or drooping ears.

malocclusion (n.). Defect in which the teeth do not close properly.

mammal (n.). Class of animal species all of which have self-regulating body temperature, hair on their bodies, and, in females, milk-producing glands.

manger (n.). Trough or box that holds feed.

mate (n., v.). Partner in reproduction; to breed.

mite (n.). Tiny parasite that can cause a scabby condition called ear canker inside rabbits' ears.

molt (v., n.). Shed fur; condition of shedding fur.

mucoid enteritis (n.). Disease, often fatal, that afflicts young rabbits; symptoms are loss of appetite, increased thirst, and jelly-like diarrhea.

mutation (n.). Change in the characteristics of an animal that can be passed on to its offspring.

nest box (n.). Box placed in pregnant doe's cage where she will give birth.

nocturnal (adj.). Animal that is active at night.

parent stock (n.). Animals used to produce offspring.

pedigree (n.). Written record of an animal's ancestors, going back at least three generations.

pelt (n.). Skin and fur of a rabbit or other animal.

pika (n.). Small mammal of the order Lagomorpha that has short ears and lives in rocky, mountain areas.

pregnant (adj.). Having a baby growing inside.

preventative (adj.).Used to prevent or ward off problems, such as disease.

purebred (adj.). Born of parents who are of the same breed.

pus (n.). Whitish liquid produced by infection.

rabbitry (n.). Place where rabbits are kept.

ration (n.). Portion for feeding.

registered (adj.). Examined and approved by an official registrar of the A.R.B.A. Only purebred rabbits with pedigrees can be registered.

registration (n.). Examination of a rabbit and recording of its pedigree by an official A.R.B.A. registrar.

rodent (n.). Mammal of the order Rodentia, that includes mice, rats, squirrels, and beavers.

sanctioned (adj.). Authorized or approved by the A.R.B.A.

sired (v.). Fathered.

sulfa drugs (n.). Group of drugs that contain sulfa as an ingredient; used to treat and prevent some rabbit diseases, such as coccidiosis.

sunstroke (n.). Heat stroke caused by exposure to the sun.

tattoo (v.). Place a permanent identification mark inside a rabbit's ear using special tools and dye.

territorial (adj.). Tending by instinct to protect or defend a certain area against intruders.

type (n.) The general physical make-up of a rabbit.

variety (n.). A subdivision of a breed based on color.

warm-blooded (adj.). Able to maintain a fairly constant, warm body temperature independent of the temperature in the environment.

wean (v.). Change a baby's way of feeding from nursing to eating other feed; separate the baby from its mother.

wool block (n.). Illness in rabbits caused by swallowed fur that forms a blockage in the digestive tract.

INDEX

A

Abscess, 61–62
Achievement contest, 138
Advertising, 113–115
Age, 16
  for breeding, 77–78
Agricultural fairs, 2
All-wire cages, 31–32
American Fuzzy Lop rabbits, 7
American Rabbit Breeders Association (A.R.B.A), *ix*, 2–3, 137–140
  registered rabbits, 130–131
  sanctioned shows, 97
Angora breeds, 6–7. *See also* Wool block
Animals, protection from, 29
Antibiotic cream, 76
  for abscess, 61–62
Apples, 50
A.R.B.A. *See* American Rabbit Breeders Association
Awards and prizes at shows, 107–108

B

Baby-saver wire, 33, 34, 37
Best Opposite winner, 106
Birth. *See* Kindling
Blindness, 89
Body type, 14, 78
Breed, *viii*, *ix*, 1–12, 99. *See also* specific breed; Standard of Perfection
  buying tips, 3, 13
Breed clubs. *See* Specialty clubs
Breeding, 5–6, 16, 77–94. *See also* Kindling
  does, 48, 78, 81
  stock for sale, 16–20, 112
  trio, 15–16
Broken pattern, 5
Buck, 15
  feeding chart, 55
Buck teeth, 68–69
Building plans
  cages, 34–46
  feeders, 51–52
  feed measure, 56–57
  hay rack, 49
  nest box, 82
  rabbit carriers, 44–46
  watering equipment, 52
Business cards, 114
Buying rabbits, 13–20, 109

C

Cabbage, 50
Cages, 29–34
Californian rabbits, 8, 11
Cardboard liners, 83
Carrots, 50
Cats, 29
Champagne D'Argent rabbits, 10
Chilling, 87
Chlorine-bleach-and-water solution, 29–30
  for coccidiosis, 65
Class, 99
Cleaning cages, 29, 91–92, 118
  and nest boxes, 88, 92
Climate, 28, 121–123
Coccidiosis, 63–66
Colds and snuffles, 66–67
Cold-weather care, 28, 82, 121–122
Colored pattern, 5
Commercial feeds, 47–48
Condition, 14, 57
  for shows, 100–101
Cooling, 71
Coop cards, 103, 105
Cooperative Extension Service, 141–143. *See also* 4-H Clubs
Corn, 50
Crocks, 50
Crossbred, 1
Culling, 93–94
Customers, 115–116
Cuterebra fly, 63

D

Daily chores, 118
Dam, 129
Defecation in newborn bunnies, 88
Dewlap, 14
Diarrhea
  caused by feed, 48, 50
  with coccidiosis, 63–66

Disease
  in damp nest boxes, 88
  problems with feed, 49
  at shows, 95
Disqualifications, 15
  when culling, 93
  before showing, 97
Doe, 15
  breeding, 78
  care when pregnant, 81, 84
  feeding, 48, 55
  after kindling, 86
  separating the litter, 92–93
Dogs, 29
Domesticated, *ix*
Doors, 33
  building plans, 38–39, 41
Double-unit all-wire hutch, 40
Dressed meat rabbits, 113
Dutch rabbits, 4

E
Ear canker, 67–68
Ears, 14
Educational contests, 139
Eliminations, 15, 93
  before showing, 97
English Angora rabbits, 6–7
Entry forms and fees for shows, 98–100, 103
Eye infections, 76, 89
Eyes, 14

F
Farm-supply stores, 30
Feed, 47–58
  for litters, 91
  when purchasing or selling a rabbit, 20, 116
Feeders, 34, 49–52, 118
  building equipment, 51–52
Feeding, 53–55, 118. *See also* Hand-feeding
  does, 58, 81, 86
  guidelines, 55, 57–58
  litters, 91
  overfeeding, 58
Feed measure, 56–58
Feet, 14. *See also* Hock
Fertilizer, *x*
Fetuses, 81
Fiber, 48
Fines, 50–51
Fingernail clippers, 69
First aid
  for abscesses, 61–62
  for chilled bunnies, 87
  for sore hocks, 70–71
  for sunstroke/heatstroke, 71–72
  after toenail trimming, 120
First-aid kit, 74–76
Floors in hutches, 32, 37, 71
Florida White rabbits, 11
Football hold, 23–24
Forefoot, 14
Formula, 87–88
Fostering, 85–86
4-H Clubs, 18, 135–137, 141–143
French Angora rabbits, 7
Fresh greens, 50
Fur, *ix*, 11–12
  in nest box, 84, 85
Furnishings (wool), 7

G
Galvanized wire, 32
Genitals, 88
  sexing the litter, 91
Giant Angora rabbits, 7
Grand Champion award, 108
Group or variety, 99

H
Hand-feeding, 87–88
Handling rabbits, 21–26, 118
  baby bunnies, 90
  when purchasing, 19
  for shows, 100
Hardware cloth, 32
Hares, *vi–viii*
Hay, 49, 50, 101
  for nest boxes, 83
Hay mangers, 49
Head, turning a rabbit over, 25
Health checklist, 60
Health of rabbits, 59–76, 118. *See also* First aid
  for breeding, 78
  when buying, 13–14
  when culling, 93
  and feeding, 54–55, 57, 58
  problems, 61–74
  after showing, 96
Heatstroke, 71–72
Hind feet, 14. *See also* Sore hocks
History of rabbits, *viii*
Hobby, raising rabbits as, *ix*, 5–6. *See also* Showing rabbits
Hocks, 14
  sore hocks, 37, 69–71
Holland Lop rabbits, 5
Hot-weather care, 102, 122–123
  sunstroke/heatstroke, 71–72
Hutch, 15, 27–46, 59, 64
  building plans, 35–44
Hutch card, 80–81, 131–132

I
Ice for heatstroke, 72
Indoor cages, 32
Infections, 61, 71
Inherit, 12
Instinct, 79

J
J-clips, 30, 37
Jersey Woolly rabbits, 7
Judging at shows, 103–106

K
Kindle, 80
Kindling, 81–84. *See also* Nest box; Pre-kindling behavior

L
Laboratory use, *x*
Lagomorphs, *vi*
Leg (award), 107

Lettuce, 50
Lifting the rabbit, 21, 22–24
Litter, 27, 84–87. *See also* Kindling
  feeding chart, 55
  health check, 119, 121
Loin, 14

M
Mail-order supplies, 30, 141
Malocclusion, 68–69
Mammals, *v–vi*
Management contest, 138
Management records, 132–134
Mate, 78
Mating, 79–80
Meat production, 8, 112–113
  breeds for, 8–11
  and medications, 66
Medication, 121
  for coccidiosis, 65–66
  in first-aid kit, 89
Metal nest box, 83
Mineral oil for mites, 67–68
Mini Lop rabbits, 5, 27
Mites, 67–68
Miticide, 67, 76
Molting, 100
Monthly chores, 119–121
Monthly management chart, 134
Multiple-unit hutch, 39–41
Mutation, 12

N
Nail clippers, 69, 75
Nape of neck, 23–24
National Youth Rabbit Show, 138
Neomycin, 89
Nest box, 82–84, 92
  cold-weather care, 121
Nest eye infections, 76
Netherland Dwarf rabbits, 4, 27
Newborn bunnies, 84–86
New Rabbit Owner's Kit, 116
New Zealand rabbits, 8, 11
Nocturnal animals, 53
Nose, 14
Nurse, 86
Nutrition, 59

O
Oats, 50
Olive oil for mites, 67–68
Opossums, 29
Overfeeding, 58

P
Palomino rabbits, 10
Papaya, 73, 76
Parasite(s), 63, 74
  due to feeding hay, 49
Parent stock, 1
Parts of a rabbit, 14
Pasteurella multocid, 66–67
Pedigree, 17, 19–20, 94, 112
  how to fill out, 128–130
Pellets, 47–48, 50
  feed measure, 56, 58
  in self-feeders, 51
Pelt, 11
Perfume, 86
Petroleum jelly, 127
Pets, *viii, ix,* 3–5, 111–112
Picking up the rabbit, 22–24
Pikas, *viii*
Pineapple, 73
Piperazine, 74, 76
Plastic water bottle, 52–53
Poultry wire, 32
Pre-kindling behavior, 84
Protozoans, 64
Purebred, 1, 16. *See also* Pedigree

R
Rabbit carriers, 44–46, 101–102
Rabbit clubs, 109
Rabbit coolers, 122
Rabbit judging contest, 108, 138–139
Rabbit Quiz bowls, 136
Rabbitry, 18–20
  management tips, 117–134
  supplies, 31, 109
Raccoons, 29
Record keeping, 120, 127–128. *See also* Hutch card; Monthly Management Chart; Pedigrees
Registration papers, 130–131
Rex rabbits, 11, 12, 23
Rodents, *viii,* 29
Royalty contest, 139
Rump, 14

S
Salt, 49
Sanctioned shows, 2, 97
Satin rabbits, 12, 23
Satin Angora rabbits, 7
Scratches (rabbit shows), 103
Second-hand cages, 29–30
Self-feeders, 50–51
Selling rabbits, 1, 111–116
  New Rabbit Owner's Kit, 116
  at shows, 95, 109
Sex, 99
Sexing the litter, 90–91
Shade, 122
Show book (show catalog), 97– 98
Showing rabbits, 2, 95–110
  entry forms and fees, 98, 103
  show courtesy, 17
Showmanship contests, 26, 108
Single hutch, 35–39
  wooden frame for, 41–44
Sire, 129
Snuffles, 66–67
Sore hocks, 14, 37, 69–71
  medication for, 76
Sources
  for cages, 30–31
  for hutch plans, 35
  for pedigree blanks, 128–129
  rabbitry supplies, 30, 141
Specialty clubs, 17–18
Standard of Perfection, 2, 14

eliminations and disqualifications, 15
judging at rabbit shows, 104
Straw, 83
Stress, 95
Substitutions when showing, 100
Sulfaquinoxaline, 76
for coccidiosis, 65–66
Sunflower seeds, 50
Sunstroke/heatstroke, 71–72
Supplies, 30, 141
hutch cards, 80–81, 131–132
pedigree forms, 128
weekly check, 118

T
Tattooing, 94, 123–127
for shows, 99, 101
Teeth, *vii*, 14
malocclusion, 68–69
Territorial instinct, 79
Thinness, 57
Toenails, 21, 119–120
Transporting rabbits, 101–102. *See also* Rabbit carriers
Treats, 50
Trio, 15–16
Turning a rabbit over, 24–26
Type, 14, 78

U
Urination in newborn bunnies, 88
Urine guards, 33–34
Uterus, 81

V
Vanilla, 86
Ventilation, 32, 60, 121–122
Veterinary care, 76

W
Water, 49, 59, 118
dishes, 52–53
after kindling, 86
and hot-weather care, 72, 122
at shows, 102
Weaning, 16, 92–93
feeding chart, 55
Weather protection, 28. *See also* Wooden frame for outdoor hutches
Weekly chores, 118–119
Wire for cages, 31–32
Wire nest box, 83
Wolf teeth, 68–69
Wooden frame for hutches, 41–44
Wood nest box, 82–83
Wool, *ix–x*, 6–7
Wool block, 72–73
Worms, 74